Jean-Martin Fortier has done it again. Incredibly beautiful, this series of programs and professional instructional manuals makes it easy for anyone to start market gardening at a professional and regenerative level with a tailored focus on each aspect.

— Matt Powers, author, *Regenerative Soil* and *Regenerative Soil Microscopy*

Applying a market gardening mindset to your home garden will improve your yields and greatly reduce the amount of work involved. To do this, look no further than this series from Jean-Martin Fortier who is a pioneer in regenerative, biointensive, and organic market gardening.

— Rob and Michelle Avis, Verge Permaculture /
5th World, co-authors, *Building Your Permaculture Property*

Praise for the Work of Jean-Martin Fortier

This is a thorough farming manual that lays out a human-scale farming system centered on good growing practices and appropriate technology. Had I read this book when I was a starting farmer, I would now be farming with a walking tractor on an acre and hailing Jean-Martin as my market gardening guru! This book is going to inspire new farmers to stay small and farm profitably.

—Dan Brisebois, farmer, Tourne-Sol Cooperative Farm, author, *The Seed Farmer*

This is a fantastic addition to any aspiring market gardener's library, and even has a few new ideas for old hands. Jean-Martin has laid out all of the basics for how we can farm more profitability, productively, and passionately on a more human-sized scale.

—Josh Volk, Slow Hand Farm, Portland, Oregon

Jean-Martin Fortier extols the virtues of being small-scale, and expertly details the use of such scale-appropriate tools as broadforks, seeders, hoes, flame weeders, low tunnels, high tunnels, and many other unique tools, specifically designed for this brand of farming. [He provides] beginning farmers a solid framework of the information they need to start up and become successful small-scale organic growers themselves.

—Adam Lemieux, Product Manager of Tools & Supplies,
Johnny's Selected Seeds

GROWER'S GUIDES
Jean-Martin Fortier
FROM THE MARKET GARDENER

Living
Soil

A Grower's Guide

TRANSLATED BY LAURIE BENNETT
EDITED BY PIERRE NESSMANN
WITH THE COLLABORATION OF PATRICIA TOUYRE
ILLUSTRATIONS BY FLORE AVRAM

new society
PUBLISHERS
www.newsociety.com

© Delachaux et Niestlé, Paris, 2024 First published in France
under the title: *Un sol vivant. Les guides du jardinier-maraîcher*
Jean-Martin Fortier, Flore Avram, Patricia Touyre

Inquiries regarding requests to reprint all or part of *Living Soil*
should be addressed to New Society Publishers at the address below.
To order directly from the publishers, please call 250-247-9737
or order online at www.newsociety.com.

Any other inquiries can be directed by mail to:
New Society Publishers P.O. Box 189, Gabriola Island, BC
V0R 1X0, Canada
(250) 247-9737

New Society Publishers is EU Compliant.
See newsociety.com for more information.

LIBRARY AND ARCHIVES CANADA CATALOGUING IN PUBLICATION
Title: Living soil: a grower's guide / Jean-Martin Fortier;
   translated by Laurie Bennett; edited by Pierre Nessmann with the
   collaboration of Patricia Touyre; illustrations by Flore Avram.
Other titles: Sol vivant. English
Names: Fortier, Jean-Martin, author | Bennett, Laurie, translator. |
   Nessmann, Pierre, editor | Touyre, Patricia, editor. | Avram, Flore, illustrator.
Description: Series statement: Grower's guides from the market gardener |
   Translation of: *Un sol vivant.*
Identifiers: Canadiana (print) 20250216957 | Canadiana (ebook)
   20250216965 | ISBN 9781774060186 (softcover) |
   ISBN 9781550928112 (PDF) | ISBN 9781771424073 (EPUB)
Subjects: LCSH: Garden soils. | LCSH: Soil management. | LCSH: Gardening.
Classification: LCC S596.75 .F6713 2025 | DDC 631.4—dc23

Funded by the Government of Canada   Financé par le gouvernement du Canada

New Society Publishers' mission is to publish books that contribute in
fundamental ways to building an ecologically sustainable and just society, and to
do so with the least possible impact on the environment, in a manner that models
this vision.

Creating a future where humans live in harmony with nature and with each other

Founded by Jean-Martin Fortier, the Market Gardener Institute is committed to inspiring and supporting new organic growers at every stage of their journey. Our mission is to equip them with the essential technical skills needed to thrive in their vital agricultural work.

Our vision is to multiply the number of organic, regenerative farms around the world and create a future where humans live in harmony with nature and each other.

www.themarketgardener.com

Presenting the collection

Grower's Guides from the Market Gardener

Hi!

I am delighted to bring you this new collection of practical guides. The advice you'll find in these books is based on working methods I developed on my own microfarm and refined over the last two decades. While plenty of these concepts are not new and were passed on to me by different mentors through the years, many other ideas stem from my own farming experience. I am sure you'll come across a number of tips and tricks that are innovative, proven, and easy to implement.

Whether you are a home gardener, hobby farmer, new market gardener, or an experienced farmer looking to transition to more intensive growing on smaller plots, you will find everything you need to take your horticultural practices even further.

Wishing you success and happiness in your agricultural adventures!

Jean-Martin Fortier, market gardener in Saint-Armand, Quebec

Contents

Introduction:
A Few Words About
My Background

Drawing on principles from agroecology, permaculture, and entrepreneurship, I champion a modern form of nonmechanized farming, carried out on a human scale.

On a human scale means feeding many local families, while respecting the human and natural ecosystems in which we operate.

On a human scale means allowing market gardeners to make a decent living from their work, to run their businesses as they see fit, and to give themselves more time off than conventional farmers.

On a human scale means evolving through the use of technology but especially by relying on people and their skills and knowledge.

From Organic Farms...
I studied agroecology at McGill University's School of Environment in Montréal, where I met my wife and business partner, Maude-Hélène Desroches. At the time, we were both looking to create a new model for farming, one that would have a positive environmental impact. After graduation, we spent two years in New Mexico, USA, working on an organic farm and learning to be market gardeners.

Our microfarming aspirations were later fueled by a trip to Cuba where we spent time on *organopónicos*, fascinating urban farms that were established during the American embargo. During that era, after the fall of the USSR, the country developed a biointensive and urban agricultural model to ensure food security for the island's residents.

... to a Family-Run Microfarm

Back in Quebec in 2004, we acquired a small plot of 10 acres in Saint-Armand, in the scenic Eastern Townships. On this land, we experimented with our innovative approach to market gardening, which especially drew from the work of Eliot Coleman, an American market gardener who has been highly influential in the world of organic microfarming.

We built a 2-acre market garden, Les Jardins de la Grelinette, where we were able to test the first iterations of my method, now called the Market Gardener Method. It consists of crop rotation, the near-exclusive use of hand tools, organic growing practices, and shorter marketing channels, with direct sales made through CSA boxes and farmers' markets. At Les Jardins de la Grelinette, Maude-Hélène and I both worked full-time, and hired two farm workers (one full-time and the other part-time) to help with harvests.

Making 2 Acres Profitable

Success came quickly, both in terms of harvests and direct sales. After bringing in $33,000 in our first year, we earned twice that in the following year, and more than $110,000 in our third year of operation.

We were thus able to earn a living as market gardeners from almost the very beginning. Since then, our farm has continued to feed more than 200 families every year, offering roughly 40 types of vegetables, all grown on just 2 acres. Over the years, our harvests expanded and sales continued to increase. Eight years after starting the farm, I presented this farming model in a practical guide called *The Market Gardener* in 2014. The book was an instant success — over 250,000 copies have now been sold, and it has been translated into nine languages.

In 2015, with the support of a generous patron, I founded Ferme des Quatre-Temps in Hemmingford, Quebec, with the vision of creating a model for the future of ecological agriculture. On this 160-acre farm, we established a polyculture system in a closed-loop cycle, raising pasture-fed cattle, pigs, and hens, alongside a culinary laboratory. At the heart of the farm, 7.5 acres were dedicated to a market garden, where we applied the growing methods developed at Les Jardins de la Grelinette. It is here that I teach my apprentices the principles of productive and profitable market gardening.

The project was featured in a TV show called *Les fermiers*, which follows the evolution of Ferme des Quatre-Temps and its apprentices, who later start their own farms in front of the cameras. The show was a hit in Quebec and is now available on TV5 Monde and Apple TV.

In parallel, I worked to expand my methods to reach a broader, global audience. In 2018, we launched the Market Gardener Masterclass, a fully online course now available in over 90 countries. To further support this initiative, I founded the Market Gardener Institute with a clear mission: to educate the next generation of growers by equipping them with the knowledge, skills, and resources needed to become leaders in the organic farming movement.

The Institute has two key objectives: to teach best practices in market gardening techniques and growing methods, and to demonstrate that small-scale farming worldwide can not only be ecological but also productive and profitable. On a global scale, it's the number of farms, not their size, that holds the key to feeding the world.

Inspiring Change

My ambition is to drive meaningful change in society by promoting a way of farming that honors nature, supports communities, and empowers local farmers. I believe in a decentralized farming model, built farm by farm, as the foundation for a truly sustainable and resilient food system.

Since 2020, I have proudly served as an ambassador for the prestigious Rodale Institute, which researches regenerative organic farming practices in the United States and beyond. I am also honored to be the ambassador for Growers and Co., a company that develops tools and apparel for new organic growers. In 2023, we launched Espace Old Mill, a restaurant and market garden set in one location. The restaurant uses the best produce in the region, including harvests from our own farm.

What Is the
Market Gardener Method?

While my approach may seem innovative, it is founded on practices that were first developed by 19th-century Parisian gardeners, who fed more than two million people through a network of thousands of market gardens—precursors to our modern-day microfarms—within the city of Paris.

These market gardeners applied remarkable ingenuity, skills, and knowledge to meet the increasing food demands of a city in the midst of urbanization and demographic expansion. They achieved this through organic, nonmechanized agriculture. From the mid-18th century to the 20th century, many books were written about the innovative practices of these market gardeners, whose technical feats were admired throughout Europe. But with the advent of modern practices, much of this know-how was relegated to the past.

As a result of mechanization, the advent of agronomic science, and improved refrigeration and transport that brought in fresh and inexpensive food grown abroad, farms grew in size, became less diversified, and took on a more technological focus—a trend that continues today.

Fortunately, these inspiring models led to the development of horticultural methods that have endured, and with the same objective: to grow sustainably, by maximizing vegetable yields without degrading soil quality. We now use the term "biointensive" to describe these methods. Unlike extensive agricultural operations, they continue to work on a human scale and offer farmers the opportunity to use little mechanization. Despite what some may believe, this approach is also profitable.

By working on only small plots of land, market gardeners can keep start-up investments to a minimum, compared to the funds needed for a conventional farm. Biointensive farmers also require a smaller workforce, doing the work themselves with the help of just a few employees. They also sell their produce directly to customers, avoiding commissions to intermediaries. These three factors allow market gardeners to start generating profits quickly.

Still, it's important to remember that working the land is never easy. While market gardeners can make a good living with this method, the first seasons are time-consuming and require a significant workload and financial investment. In this profession, nothing comes easy, and every dollar you earn is the fruit of your labor, the result of your organizational skills. That's why I always tell my apprentices to learn how to work smarter, not harder.

From a financial perspective, market gardeners should plan to start with an investment of $50,000 to $150,000, depending on whether certain assets are already available—such as a building that can be converted, access to abundant water, electricity, natural gas, or a vehicle. This amount does not include the cost of purchasing land, which can be amortized over 20 years, if needed. Renting is also an option that can prove very profitable, especially when the farm is located near a city or an affluent municipality, where land is expensive.

Regardless of experience and preparation, the first years of market gardening will be intense. Opening new ground, constructing greenhouses and tunnels, and setting up infrastructure (irrigation, washing and packing stations, nurseries, etc.) all take extra time and effort. However, once this phase is complete, market gardeners who have mastered their craft can do more than just make a living off a few acres—they can earn a very decent living.

This leads to another key principle I teach: your farm should work for you, not the other way around. Profitable and productive farming is possible, but you need to set it up for success.

Preface:
Healthy Soil,
Healthy Vegetables

For over two decades now, I have dedicated myself to biointensive agriculture, where preserving life in the soil is a major factor leading to crop success. I started with small plots of land that I transformed into living labs, and my journey led me to explore the countless mysteries that lie beneath our feet, to study nature's cycles, and to learn what the soil can teach us. It is this very expertise—experience built over time and inspired by nature—that I want to share with you in this guide. As I tended the soil and delved into its inner workings, I learned to take a close look at nature, to understand subtle signs made by plants, insects, and microorganisms. I discovered a complex ecosystem, a remarkable community made up of billions of microorganisms, bacteria, fungi, and earthworms working in harmony to support the plant world. The way I see it, the more life there is in the soil, the healthier and more fertile it is. Respect soil life, feed it with rich organic matter, and cultivate the earth without traumatizing it. As I applied these principles, my soils became darker—a sign of increased organic matter—but, more importantly, they evolved to become more alive and thus more fertile.

Through the pages of this guide, join me and discover the fundamental principles of growing vegetables in living soils. We will take a deep dive into the world of microorganisms that thrive underground and learn how to understand and cultivate the land in a way that is gentler, more measured, and, especially, that respects these ecosystems.

May this journey to the heart of living soils be a source of inspiration and hope for you, just as it once was for me. Going forward, together, let's cultivate soil life and, in so doing, cultivate a better future.

Jean-Martin Fortier, market gardener in Saint-Armand, Quebec

The Living Soil

Our biointensive market gardening system is primarily based on creating and maintaining living soils, which promote plant growth and, ultimately, fruiting—all while increasing fertility and biodiversity.

Soil: One of Three Fundamental Environments for Plants

Plants live at the intersection of three natural environments: the lithosphere (soil, ground), the atmosphere (air), and the local ecosystem (biological world).

The soil, underground, is where plant roots grow. It serves multiple purposes, including acting as a pantry to supply plants with the nutrients they need. It also provides physical support, firmly anchoring plants in the soil, which is especially relevant for fruit-bearing vegetables that tend to grow taller and are generally bigger.

The atmosphere influences the above ground portion of the plant, which is constantly exposed to climatic conditions, which make all plant life possible. It delivers sunlight for photosynthesis, provides atmospheric gases such as carbon dioxide, and drives soil formation and evolution through weathering of rock and the Earth's crust.

The local ecosystem consists of all living organisms, plant and animal that come into contact with the plant. These relationships can be beneficial or detrimental to it.

The three environments work in concert to satisfy the needs of plants, both physiological (water, oxygen, light, and heat) and nutritional (water, CO_2, mineral and organic elements).

To continuously improve soil quality, the Market Gardener Method applies the following principles: minimum tillage, standardized cultivation systems, and an approach that is both organic and intensive. Above all, growers' practices should be based on a deep understanding of their own soil.

Minimum Tillage

The Market Gardener Method relies on shallow tillage, in only the top 2 to 4 inches (5–10 cm) of soil, leaving most soil layers undisturbed. Because vegetable microfarms operate without large mechanized agricultural machinery they can optimize growing space. In other words, if you don't need a turnaround area for tractors, you need less space between beds, and if those beds will not have to accommodate large machinery, then rows can be sown closer together. The benefits here are twofold: growers can optimize their small market-gardening plots, while also limiting the costs associated with the purchase, use, and maintenance of a large fleet of agricultural machinery. Another notable advantage is avoiding soil compaction, which hinders crop development. However, a tractor may be essential to complete specific tasks like moving heavy loads, stirring and turning compost, and carrying equipment or materials around the farm. So, while a profitable microfarm can be successfully run without a tractor, the Market Gardener Method does not strictly prohibit its use.

Biointensive market gardeners should prioritize equipment, such as tilthers, broadforks, and various hand tools to cultivate soil, as well as practices including occultation (tarping) and, especially, permanent beds. These processes only lightly disturb the soil, protecting its organic matter while suppressing weeds, thereby encouraging the development of crops and beneficial soil organisms. Maintaining a ground cover—with either a cover crop or a silage tarp—keeps soil moisture in a healthy range and limits nutrient leaching and soil erosion caused by precipitation and irrigation. For more on appropriate tools and their uses, see *Vegetable Garden Tools* by Jean-Martin Fortier.

Standardized Systems

When operations and garden plots are more standardized, efficiency increases and growers save time. Therefore, it all starts with creating permanent beds and aisles grouped into predefined blocks. Growers can then calibrate all their equipment to fit the dimensions of these garden beds. Crops must also be carefully planned, from crop rotations and successions within a bed, to farm tasks scheduled for certain times of the year. Armed with this detailed calendar, growers can efficiently allocate the operations to be completed from week to week and even from one day to the next.

Cultivating the same beds every season is the simplest way to improve soil quality. It limits compaction in the beds and reduces the need for amendments, as inputs can be used more effectively when they are not spread in the aisles. If you are dealing with cold weather and heavy soils, we recommend raising beds by a few inches to promote natural drainage and to help the soil warm up faster in the spring.

Biointensive Farming

Lastly, to improve yields, plant vegetables quite densely, with rows set as close together as possible. This enhances productivity, facilitates harvesting by hand, and streamlines the use of equipment like irrigation systems, tarps, row covers, and insect netting. Crops grown biointensively also provide ground cover quickly, preventing weeds from germinating and developing.

This results in increased yields as beds can be cultivated multiple times through the use of crop successions. Depending on your regional climate and available equipment, such as greenhouses or tunnels, you can grow one to four—or even five—crop successions per year in a single bed. To apply this principle, you must adjust your planting schedule accordingly. When each crop has an allocated spot in the farm calendar, you optimize the use of every bed. If you carefully stick to the plan, they will never be empty for long, leaving little space and time for weeds to grow. With this approach, the soil is exclusively dedicated to growing vegetables and will deliver maximum yields.

Understanding Your Soil: Foundations for a Good System

The Market Gardener Method is particularly well suited to developing small-scale market gardening systems that deliver high yields. Primarily, it relies on growers having a good understanding of their soils, so they can work them wisely. To succeed, you need to ask yourself the right questions about your soil type and, above all, use common sense to answer them! What is it made up of? How can you track changes in the soil? How can you improve it? How can you maintain fertility and stimulate soil life? These are the main themes addressed in this book.

All organic vegetable growers, whether professionals or home gardeners, must be familiar with their soil, its composition, and the interactions between all organisms within it. Under their feet lies a full-fledged organic waste recycling and recovery facility that runs day and night. With the help of various insects, animals, and microorganisms, such as bacteria, algae, and fungi, organic waste is transformed into nutrients that are essential to plant life. The top few inches of soil, where plant roots grow, are the most biologically active. This layer is called the rhizosphere. Market gardeners who respect their soil and strive to improve it—especially when based on an analysis that identifies amendment needs—can see a significant increase in yields if they are careful not to exhaust the soil. When the soil is supported by interactions between fauna and flora, and given the required attention, it can become a thriving ecosystem in its own right—a living soil. But how can you get to know your soil and thus improve your vegetable garden or farm?

What Is a Living Soil?

What Are Soils Made Up Of?

Soil is constantly evolving. It builds structure and breaks down in response to myriad external events and other factors. It is alive, so changes occurring within it—both positive and negative—depend on factors such as climate and human actions. From the parent rock through to the surface of the earth, soil forms multiple layers that vary depending on its mineral content and location. Soil is shaped by intense plant, animal, and microbial activity and the ecosystem that supports it. It is therefore a living substrate comprised of organic matter needed to sustain plant, animal, and microbial life, as well as minerals, liquids, and gases. After you identify and understand these components, their ratios, and the type of life that thrives in your soil, you will be better equipped to maintain, enrich, and prepare it for future crops.

Soil Layers

―――――――○―――――――

By looking at the layers that make up your soil, you can determine their composition, assess their biological activity, ascertain any mineral loss through leaching, and locate root systems and rocks.

Soil Horizons

In pedology—the science of soil formation and evolution—the word "soil" refers to the ground between the surface and the parent rock, which is the upper mineral layer in the Earth's crust. Soils form as the rock weathers, creating a soil profile made up of distinct layers, called horizons, that are roughly parallel to the surface. Each horizon is symbolized by letters or numbers and has its own visual characteristics and morphological properties:

· O for "organic." This is the top horizon, just below the surface, and is made up of organic matter.

· A for "arable." This horizon contains both minerals and organic matter.

· B. This layer tends to have more color and is where leached minerals accumulate. Market gardeners commonly refer to it as the subsoil.

· C. This is the deepest horizon. It lies just above the parent rock and is made up of partially weathered parent material.

Sometimes you can find a horizon labeled E for "eluvial." This lighter-colored layer, which is not always present in the soil profile, develops as a result of leaching. In the field, the easiest way to view these layers is to make a deep, vertical cut in the soil, about 20 inches (50 cm) or more depending on your soil depth. This is called a soil profile.

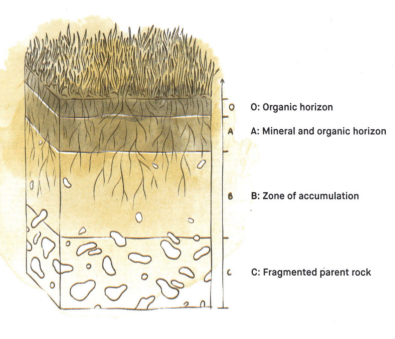

O: Organic horizon

A: Mineral and organic horizon

B: Zone of accumulation

C: Fragmented parent rock

STANDARD SOIL PROFILE AND BRIEF NOMENCLATURE
The layers containing humus are the most fertile and can be recognized by their near-black color. They form under the litter, the uppermost soil layer, which primarily consists of undecomposed plant residues like dead leaves, twigs, and clumps of dried grass.

Note from Jean-Martin Fortier

Although market gardeners and home gardeners may only be concerned with the soil layers they cultivate, to a depth of about 12 inches (30 cm), it's also important to focus on the lower layers, the subsoil, where root systems grow. This area is a precious source of nutrients, a kind of pantry for plants. You can feed it by growing cover crops and leaving the deep roots of some vegetables in the ground where they can gradually decompose.

Soil Evolution

The age of a soil typically correlates to its maturity. Forest soils are considered mature, while cultivated soils are deemed immature as their natural evolution is constantly interrupted by human intervention.

Parent Rock

The rock found in the deepest soil layer is the starting point for pedogenesis, or soil formation. In a weathering process shaped by climate and pioneering species (flora), mostly mosses and lichens, the parent rock gradually breaks down. This soil formation can be interrupted by factors related to climate, such as erosion or leaching, as well as geological events and human activity.

Anthroposols

Some soils, called anthroposols, have been profoundly impacted by humans, especially by farming practices. They undergo near-constant alterations that interrupt pedogenesis. When soils are continually tilled at varying depths, their horizons are regularly turned over, stirred, and mixed. Organic matter cannot gradually build up and become stable because the process is perpetually disturbed by external inputs, both organic and inorganic.

In these very young or immature soils, the soil profile is simplified, featuring only two horizons: horizon A contains the organic matter that the market gardener has incorporated into the soil, and horizon C generally consists of a thick layer of silt. Horizon O is practically nonexistent. On many cultivated plots, all plant matter, including unused parts (roots, stems, leaves), are removed in the harvest and tilling process. As a result, the soil is unable to turn decomposing crop residues into humus. Naturally occurring soil colloids (see p. 27) are altered and eventually disappear, and the soil gradually becomes a nonliving environment.

SOIL FORMATION

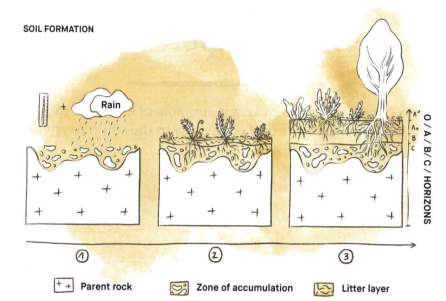

O / A / B / C / HORIZONS

Parent rock · Zone of accumulation · Litter layer

1. The parent rock breaks down due to the mechanical action of cold temperatures and dissolves by the chemical weathering of water. This process creates a B horizon in which organic and inorganic materials accumulate.

2. Pioneering species accelerate chemical weathering in the parent rock material and generate organic matter, which forms litter and humic substances.

3. Litter and humus develop, in different horizons, and various plants begin to grow.

To top it all off, heavy motorized equipment is often driven over these plots, further compacting the soil. Hence the appeal of the Market Gardener Method; it advocates working the soil manually to avoid compaction and retaining the roots of certain crops after harvest to restore organic matter deep in the soil.

This method is inspired by soil in nature that feeds on organic matter that is first deposited onto the litter, the surface layer covering the soil. It is made up of residues from plants (leaves, twigs, etc.), fungi (spores and mycelia), and animals (excrements, invertebrate remains, etc.).

Soil Composition

Soil horizons contain various components, such as gases, liquids, and minerals that contribute to the soil ecosystem.

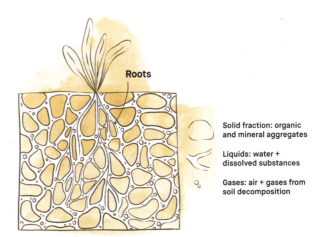

Roots

SOIL COMPOSITION
Soils are composed of solid fractions—aggregates—as well as liquids and gases that fill the gaps (pores) between them. Roots also make their way into these pores.

Solid fraction: organic and mineral aggregates

Liquids: water + dissolved substances

Gases: air + gases from soil decomposition

What Are Soils Made Up Of?

Soils consist of components that exist in different states. The gaseous content includes gases that are identical to those found in the atmosphere, such as carbon dioxide and oxygen, as well as gaseous byproducts of decaying organic matter, such as methane. The liquid component, referred to as the soil solution, contains water and ions that allow it to dissolve soluble material, including soil minerals, organic matter, and amendments. This combination of water and dissolved nutrients is a source of essential nutrients for plants. The solid content (organic and mineral matter), the largest by volume, determines the texture and nature of the soil. These components combine to create soil, which is an ecosystem in itself.

Soil Mineral Content

Soil minerals are remnants of weathered parent rock. Over time, several phenomena transform this material, such as temperature changes that cause repeated expansion and contraction. Water infiltration, in small or large amounts, is a significant factor contributing to erosion. Wind also causes erosion and, like runoff, adds minerals that can modify soil nature. Parent rock determines the type of soil found in a region, and thus in your own farm or garden. Soils that form over sand, sandstone, and granite tend to be acidic, while those borne from limestone or chalk are more alkaline. The size of the mineral particles in soils depends on the type of weathering undergone by the parent rock. Particle size offers a way to classify soils by distinguishing between coarse fractions and fine earth. You can also identify soil texture (the percentage of each mineral type) to determine whether your soil has more clay, loam, or sand.

Particle Size

Soil tests provide a range of information and consist of both particle size analyses and chemical analyses, where the former indicates the percentage and size of mineral particles in the soil.

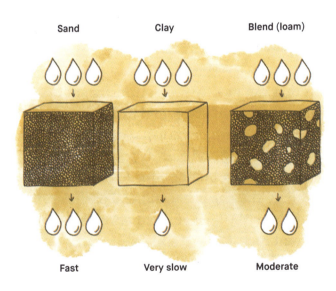

WATER INFILTRATION & DRAINAGE ACCORDING TO PARTICLE SIZE
In sandy soils, rain and irrigation water drain fast and thoroughly, whereas it stagnates and passes more slowly through clay soils. In balanced soils (called loamy), which combine sand, clay, and aggregates (gravel and coarse sand), some water is retained, while the rest percolates deeper into the ground, in moderate amounts. This is more beneficial to plants.

To determine the percentage of each mineral type in your soil, here is a quick particle size test. Drop a small amount of soil into a clear tube or straight-sided jar, add distilled water, and shake or stir. After sitting for a minute, the sand will settle (A tubes). After 24 hours, the silt content will have settled onto the sand (B tubes). The water will remain relatively colored due to suspended clay content, made up of colloidal micelles (see page 27) that are too small to settle.

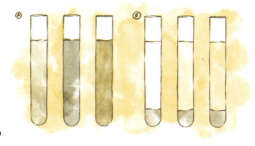

Coarse Soil

Coarse soils are made up of larger soil minerals that form what is sometimes called the soil skeleton. After sifting a soil sample, you may find coarse fragments like cobbles (more than 5 cm in diameter), medium and coarse gravel (5 mm to 5 cm), and fine gravel (2–5 mm). This type of soil is relatively poor in terms of fertility because it does not retain nutrients. Coarse fractions abrade the metal blades of tools used to work the soil, wearing them out more quickly. In addition, coarse soil increases permeability and promotes natural drainage, which reduces capacity to hold water and organic matter, leading to nutrient leaching. Under certain conditions, pebbles and stones on the soil surface can store heat during the day and radiate it overnight. While this phenomenon is not particularly well suited to market gardening, it is useful in some vineyards.

Fine Earth

Fine earth soil analyses are conducted in a lab using a sedimentation technique that generates layers of identical particles. These layers form one above the other according to particle size, with sand at the bottom and clay at the top:

· Clay: less than 2 microns.
· Silt: 0.05–0.002 mm (or 50–2 microns).
· Sand: from 0.05 mm (or 50 microns) to more than 2 mm.

PROPERTIES OF FINE EARTH

This table summarizes the advantages and disadvantages of fine earth types.

TYPE OF FINE EARTH	IMPACT ON FERTILITY	ADVANTAGES FOR SOIL	DISADVANTAGES FOR SOIL
COARSE SAND (diameter > 2 mm)	None	- Improves air circulation and water infiltration. - Improves heat exchanges with the atmosphere, which warms the soil. - Easier to till and cultivate. - Improves root penetration.	- Poor capacity to hold water. - Increases leaching. - Reduces structural stability. - Wears out tools used for cultivation and tillage.
FINE SAND (diameter 2 to 0.05 mm [50 microns])	None	- Contributes to air circulation and water infiltration.	- More likely to become compacted. - Does not retain essential plant nutrients. - Tends to cause soil asphyxiation.
SILT (diameter 2 to 50 microns)	Low		- Leads to soil impermeability. - Causes soil asphyxiation. - When in high concentrations, soil structure is made unstable (slaking forms a soil crust).
CLAY (diameter less than 2 microns)	None	- Improves water holding capacity. - Increases soil cohesion.	-Inhibits root penetration. -Makes tillage more difficult. -Slows down heat exchanges between the atmosphere and the ground (thermal inertia). The ground is colder in the spring, but more moderate in the fall. - Reduces soil permeability and aeration.

Texture

Soil texture defines the ratios of mineral particles in the soil, classified solely according to their size, without considering their composition. The three main types are sand, silt, and clay; loam is a mix of the three.

To assess your own soil texture, see the simple experiment described on page 45.

SANDY SOILS are permeable, with good aeration and drainage. They do not compact much and are easy to cultivate and weed. However, this type of soil is prone to leaching and erosion. It tends to be less fertile because it retains little to no nutrients, which filter through the soil along with rain or irrigation water. Many plants struggle to get established because their roots have trouble accessing the water and nutrients they need, and the sandy soil provides insufficient anchoring to support foliar growth. However, root vegetables such as carrots, radishes, celeriac, and potatoes thrive in this loose and aerated soil. To keep plants from drying out, these soils require regular irrigation. Frequent amendments (compost) can mitigate the effects of leaching, while growing cover crops helps increase soil organic matter.

SILTY SOILS are formed by accumulated alluvial deposits. Made up of very fine mineral debris, they are soft to the touch. In rainy weather, this type of soil quickly becomes compacted, making it difficult for plant roots to penetrate. Those that manage to thrive may asphyxiate because the gaps between the aggregates are so small, or even nonexistent. Some vegetables, such as leek, cabbage, beans, eggplant, lettuce, beet, tomato, and melon, tolerate and even grow well in silty soils.

CLAY SOILS are generally hard to till and cultivate because they are heavy and slow to dry. Although rich in nutrients and trace elements, they are often compacted and poorly aerated, making them inhospitable for earthworms and burrowing animals. Some vegetables can adapt to clay soils and tolerate temporary waterlogging, including rhubarb, cabbage, chicory, and even tomatoes.

LOAM SOILS contain sand, silt, and clay in roughly equal proportions that balance out the stickiness and nonadhesive characteristics of the different particles. They retain moisture and nutrients, drain well, offer solid anchoring for roots, and are generally good soils for cultivation.

Soil Colloids

Colloids are nature's sponges, attracting bioelements that provide nutrients to plants. These particles are the reason soils are so special.

Minerals that have undergone significant chemical changes since breaking off from the parent rock include clays, such as kaolinite or montmorillonite, hydrated silica, aluminum oxide, iron oxide, quartz debris, and limestone crystals. In terms of particle size, they fall under the clay fraction, with a diameter of no more than 2 microns. They are made up of extremely fine aggregates called micelles, which do not dissolve. These remain enclosed in a liquid and form a paste, almost like a glue, known as a colloid.

Decomposed organic matter, or humus, is also a colloid. These two kinds of colloids are negatively charged, and just like sponges, they attract positively charged nutrients and trace elements as exchangeable bases.

Together, they form the clay-humus complex (see p. 39), which is so powerful that it attracts and holds the necessary nutrients for plants, thereby serving a critical role in soil fertility.

The Texture Triangle

To make the soil classification system easier to understand, and to get a better sense of soil types, the results of the various tests are graphically represented in the texture triangle. After your soil is analyzed, you should receive a copy of this triangle in the report.

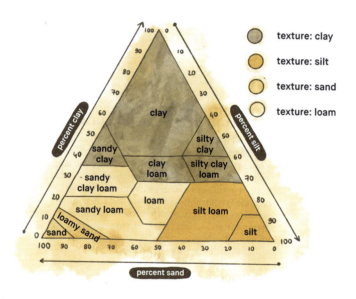

You can use the precise results of a soil analysis and this triangle to determine the particle size distribution in your soil. Connect the percentages for each mineral to identify the texture designation that characterizes your soil.

How to Interpret the Texture Triangle

Every side of the triangle features a graduated scale, ranging from 0% to 100%, and each one represents a different mineral: clay, silt, or sand. To interpret your soil sample analysis, apply the percentage of each particle type to the corresponding side of the triangle. The three lines will point to a texture designation in the triangle, indicating your soil type.

A Simpler Classification

All the precise, detailed numerical data featured in the texture triangle are useful for improving soil on a farm or in a home vegetable garden. However, for market gardeners and home gardeners alike, a simpler reading of the graph may be sufficient.

This approach classifies soils according to their clay percentage:
· Light soils: less than 10% clay
· Light-to-medium soils: from 10% to 15% clay
· Medium soils: from 15% to 25% clay
· Medium-to-heavy soils: from 25% to 30% clay
· Heavy soils: more than 30% clay.

The ideal soil for market gardening is in the medium soils category, but clay content should not exceed 20%, which may limit soil biological activity.

Note from Jean-Martin Fortier

Clay soil has a bad reputation with both home gardeners and market gardeners because it drains poorly, is hard to cultivate, and can be slow to warm up. It does, however, have the ability to hold nutrients and water (irrigation or precipitation), which is a boon for organic market gardeners looking to reduce irrigation. Working with clay soil does require some adjustments: use compost inputs to improve soil structure, apply mulches to limit damage (surface crust) from precipitation and irrigation beating down the soil, and install effective drainage systems. By changing your clay soil this way, you will establish long-lasting fertility.

To improve soil characteristics, the best strategy is to cultivate beds in the fall then loosen and aerate them again with a broadfork before the next planting. Regularly incorporating organic matter and mineral content (sand, perlite) will improve your soil over the course of several years. Finally, you may occasionally add perlite, coarse sand, or gravel in the holes or furrows to improve the soil before seeding or planting a specific crop.

Soil Organic Matter

In soils, organic matter is present in three forms: living organic matter, active organic matter, and stable organic matter (humus). They are sometimes referred to, respectively, as living, dead, and very dead. Living organic matter is remarkably biodiverse, consisting of flora—bacteria and fungi, for the most part—and fauna, which includes protozoa, nematodes, and earthworms. Active organic matter is made up of residues from both plants and animals (manure, droppings, remains). They break down easily, contributing to the stable soil organic matter, i.e., stable humus, which is essential for plant growth. Together these three types form the total soil organic matter, which contains carbon, oxygen, hydrogen, and nitrogen, as well as metals, metalloids, and trace elements. It's a real melting pot! When it becomes flocculated, i.e., when suspended, undissolved solid matter in liquid agglomerates to form larger aggregates, it behaves like cement, bonding soil mineral particles. In this state, the organic matter is referred to as colloidal.

Soil Biodiversity

Soil fauna and flora play a major role in decomposing plant detritus and animal waste, hence their name: decomposers. They work in concert within the top few inches of soil.

Soil Fauna

Sometimes called pedofauna, this category comprises macrofauna and microfauna. The former includes the well-known earthworms as well as insect larvae, pill bugs (woodlouse), and gastropods. Microfauna—at less than 2 mm in size—refers to nematodes (roundworms), springtails, and mites. These tiny soil organisms are easy to spot because they leave behind visible signs on plants. For example, springtails and pot worms (Enchytraeidae), small colorless worms, nibble on the tender parts of leaves without touching the more rigid veins.

In soils, earthworms are critical detritivores, true allies of market gardeners and home gardeners. They are divided into three ecological groups, according to the depth at which they live.

EPIGEIC EARTHWORMS are dark red and live at the soil surface, in the litter made up of leaves and other plant material. They are 1 to 5 inches (3–12 cm) long and have a short lifespan (1 to 2 months).

ENDOGEIC EARTHWORMS often have no pigmentation and have a pink or whitish sheen. They burrow horizontally down to a about 8 inches (20 cm), ingesting soil and assimilating some of its organic matter, which they cast, partially filling these tunnels. They are 1.5 to 5.5 inches (4–14 cm) long and live for 3 to 5 years.

Leaf skeletonized by springtails (small jumping insects) or by pot worms (small colorless worms).

ANECIC EARTHWORMS make vertical burrows up to 3 feet (1 m) deep. They are 2 to 10 inches (6–24 cm) long and may be brown, red, or gray. Every night, they rise to the surface to feed on organic matter then produce piles of castings, excretions that look like nutrient-rich soil. These are particularly high in nitrogen and easily assimilated by plants. As they descend into the soil, anecic earthworms carry along organic residues. Their movement creates burrows, aerating the soil and supporting microfauna and microflora, which contribute to humus formation.

Soil Flora

Soil flora refers to plant life in the soil, particularly bacteria and fungi. Two acres can contain between 45 and 4,500 pounds (20–2,000 kg) of bacteria, but not all of them are beneficial for farmers and gardeners. Some need oxygen to survive and thrive (aerobic bacteria). These ones are involved in converting organic matter into humus, giving off a pleasant earthy scent. Others exist in environments deprived of air (anaerobic bacteria) and are harmful to soils because they transform organic matter into products such as methane and hydrogen sulfide that are toxic to plants. These bacteria give off the characteristic foul odor of sulfur. Bacteria can draw upon different sources for energy. When they feed on organic matter, they are referred to as heterotrophs. Those that consume minerals but derive energy from another chemical source are called autotrophs.

Three types of earthworms and their habitats.

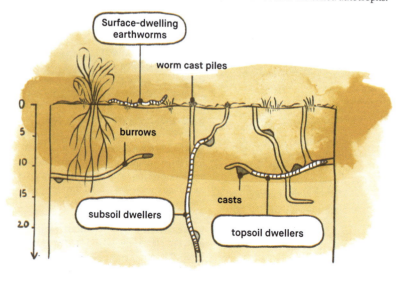

An example of *Streptomyces*. Viewed under a microscope, they have a filamentous form.

Fungal filaments (mycelia) grow around and into roots, providing many nutrients, such as nitrogen (N), phosphorus (P), and potassium (K). They also carry water.

Focus on Bacteria

Bacteria perform very specific tasks in the soil and play a critical role in the process of humus formation, known as humification. Although they live in all types of soil, bacteria prefer rich soils. Some, like rhizobia, live symbiotically with legumes, not unlike mycorrhizae. They develop small growths called nodules on the roots that help fix atmospheric nitrogen. Other types of bacteria support soil ammonification and nitrification. The former is the process by which organic nitrogen is converted into ammonia, while nitrification is the conversion of ammonium into nitrite and then nitrate. Both play a role in the sulfur and iron cycles. They are easily identified with your nose because these processes give off the earthy smell of wet soil and humus, typical in forests. These bacteria, called streptomyces, are ubiquitous in soils. Although prolific decomposers, they may also slow down some microorganic and fungal activity due to their antiseptic properties.

Bacteria play a critical role in breaking down many types of organic polymers, such as cellulose, lignin, and starch, and in forming humic substances. They also serve as food for some of the soil microfauna.

Focus on Fungi

Fungi play a critical role in organic matter decomposition and soil fertilization. Relying on a network of thread-like roots called mycelium, they form active, dense fungal colonies that help break down hard-to-digest organic matter like cellulose and lignin, thus playing an important role in decomposition and the development of fertile soil.

This mycelial network is also fundamental for a large number of plants that it has a symbiotic relationship with: the plant gives the fungus sugars produced through photosynthesis, and the fungus, through the mycelial network, provides the plant with water and nutrients. This association between plants and fungi is referred to as mycorrhizae.

Types of Mycorrhizae

 1 No mycorrhizae: in vegetable gardens, only plants in the cabbage, turnip, and radish family (Brassicaceae), and the spinach and beet family (Amaranthaceae) do not rely on fungi for support.

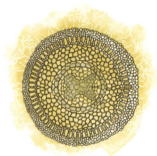

 2 Ectomycorrhiza: the mycelium forms a sheath around the fine roots of plants. This type of mycorrhization mainly affects trees in temperate regions.

3 Endomycorrhiza: the mycelium grows inside plant cells. A microscope may be required to see it.

This network is made up of underground threads that carry water, nutrients, and minerals between the plant and the fungi and from one plant to another. Fungi require an abundance of available soil organic matter to build this precious mycelial network so keep tillage light to avoid destroying it. Likewise you should also make sure to always have plants growing in the soil to continuously nourish the network, which also secretes antibiotic substances that detoxify the soil and promote plant growth.

Myxomycetes (slime molds) are single-celled organisms that are neither plants nor animals, not even fungi. They thrive on dead wood and ramial chipped wood (RCW) used in market gardening and gardening in general. They break down cellulose and feed on bacteria through phagocytosis, by engulfing organic matter. Slime molds look like relatively compact gelatinous masses before transforming into sporangia, round or elongated clusters that produce spores. But beware: some slime molds are harmful to certain crops, like cruciferous vegetables, in which they cause clubroot.

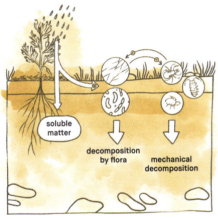

In the first couple of inches of soil, organic matter is slowly broken down by fauna and flora. This is a living soil, something that every market gardener must create and maintain on their farm.

Tip from Jean-Martin Fortier

Fungi and mycorrhizae, the unsung heroes of the soil, form an underground network that is not unlike the Internet. They connect plants by conveying water and nutrients. They are also effective protectors because they exude substances that help our vegetable crops grow bigger—and it's all natural. You can help low-fertility soils by inoculating them with mycorrhizal fungi, which are sold in a powder form to be applied when watering. This deploys a new wave of mycorrhizae in the soil and boosts plant growth that, in turn, feeds the mycorrhizae. It's a win-win interaction that brings your garden to life.

Humus: Stabilized Organic Matter

Soil flora and fauna break down living organic matter and active organic matter, gradually turning them into humic substances. This is how humus is formed, in a process called humification.

Mineralization and Humification

The component of soil organic matter that is made up of plant detritus undergoes two different transformations. Plant matter first goes through a rapid mineralization process that produces soluble or gaseous minerals (nutrients), such as potassium (K^+) and calcium ($Ca2^+$), as well as many trace elements like boron, cadmium, chromium, copper, nickel, lead, and zinc.

Then during humification, the lignified organic matter, which is much more rigid, will eventually convert into humus. In this process, bacteria and fungi break down the complex chemical molecules of cellulose, pectinate, and lignin, transforming them into humic substances.

Microorganisms break apart the long chemical molecules of lignin and cellulose into shorter molecules, creating humic substances.

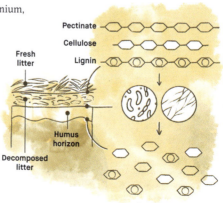

Pectinate
Cellulose
Fresh litter
Lignin
Humus horizon
Decomposed litter

Humic substances

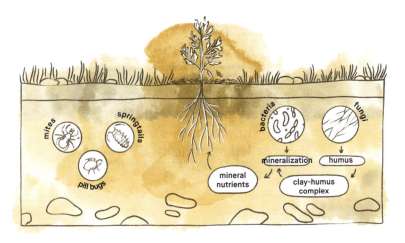

Humus forms from plant residues with the help of many living organisms.

Humus (Humified Organic Matter)

Once organic matter is humified, it aggregates with inorganic colloids in a process that is supported by earthworm activity. This is how humus (an organic colloidal complex) is created. It becomes the soil's reservoir for both nitrogen, which is mineralized by microorganisms, and other mineral nutrients. Pound for pound, humus can hold 20 to 30 times more nutrients than clay! The humus-forming process releases both carbon dioxide and nutrients that are essential for plant growth. Weather and climate can also affect the speed at which humus is formed.

Types of Humus

Humus is a general term that describes a group of humic acids. While only plant detritus can be converted into humus, animal residues can be transformed into sources of nitrogen that can be useful for soils.

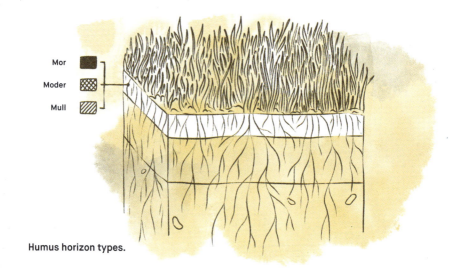

Mor

Moder

Mull

Humus horizon types.

Humus layers may contain three types of humus, depending on how rich the soil is:
· Mor is a distinct and black horizon, characteristic of poor soils.
· Moder is a blackish horizon, typical of moderately rich soils.
· Mull is an indistinct horizon that mixes into the horizon below it, indicating rich soils.

Like non-lignified organic matter, humus can also undergo mineralization, producing magnesium (Mg) or sodium (Na^+), for example, but this process is slower and depends on the soil's biological activity. Humus acts as a reservoir for soil nitrogen, containing about 5% of this key nutrient that helps plants grow big and strong. It also stores many other nutrients, far more than clay, and retains the water that vegetables need. Moreover, humus is home to all types of microorganisms, invisible laborers that bring soils to life. That's why it's so important to pamper your soil: rich soils hold the promise of a bountiful vegetable garden.

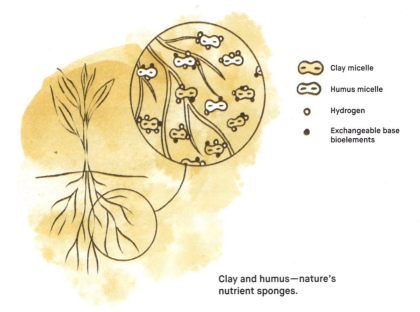

Clay micelle

Humus micelle

Hydrogen

Exchangeable base bioelements

Clay and humus—nature's nutrient sponges.

The Role of Calcium

Clay and humus form negatively charged colloids that attract positively charged particles, such as bioelements. Thanks to the calcium in the soil, these two colloids aggregate to form a clay-humus complex, the foundation of living soil. This calcium is a product of weathered limestone-rich parent rock. In the soil, it acts in three ways: it plays a physical role, promoting aggregation between clay and humus particles; it plays a chemical role, stabilizing soil pH; and it plays a biological role, improving air circulation and water penetration, while supporting microbial life. Calcium is also a critical nutrient for plants as it strengthens plant tissues, supports uniform root development, and improves fruit and seed development.

Note from Jean-Martin Fortier

Organic fertilizers and compost are important, of course, but what really matters to our vegetables is the clay-humus complex. It is the orchestra conductor guiding your soil, giving it structure, bringing it to life, and helping plants grow well and harmoniously.

Understanding your soil means learning about the type of soil on your vegetable farm or in your garden. That means the soil in which you grow crops, or plan to do so, specifically vegetables.

Understanding your soil means learning about its composition and the distribution of these components. What is the soil organic matter content? What is the soil texture and structure?

To answer these questions, you can do a few relatively simple at-home tests to get a general idea. To confirm the answers and obtain more accurate results, you may need to have a soil analysis done professionally. A lab report will help you improve soil quality on your property, which will better prepare it for future crops and improve yields—all while maintaining a healthy garden.

Understanding Your Soil

Field Tests and At-Home Experiments

Before taking any action, it's important to observe your garden or farm environment, analyzing its topography and vegetation. The type of vegetation will provide your first clues: some species develop only in soils that are basic or acidic, rich or poor, light or heavy. Next, stroll through the area to see whether the ground under your feet is soft and loose, compact and hard, dry or waterlogged. Then grab a handful of soil and knead it between your fingers to get a better sense of its nature. You don't need to be a scientist to make this initial assessment; it's largely a matter of common sense. This way you can get to know the different soils and their zones throughout your property.

The second step involves conducting a few quick experiments that will confirm your assumptions based on your initial observations and serve as a starting point for affordably monitoring the evolution of soils on your property.

Estimate Soil Fertility

The color test is quite simple and, based on your observations, will reveal how much organic and mineral matter are in your soil.

You can run this test to better understand your soil in its current form and to get a sense of its history, including previous cultivation practices undertaken on it. Over the years, you can repeat the test in a few clearly identified areas then record your observations in a notebook to keep track of the results and monitor their evolution. This test also provides a general idea of the soil texture; when rubbing the soil under your fingers, you should be able to feel crystals, mineral residues, and plant fragments.

The Method

 1 Take a pinch of soil and drop it onto a white sheet of paper, then rub the material in a circular motion under your fingertips.

2 Observe and analyze the smeared sample:

→ A black color indicates high levels of organic matter, either naturally occurring or resulting from regular soil amendments. Such soils have good potential for cultivation.

→ A beige color indicates a light soil that is regularly leached by rain. Organic matter tends to percolate into its deeper horizons. It's a relatively poor soil.

→ A rusty color, ranging from orange-yellow to reddish-brown, indicates the presence of iron in its oxidized form, meaning the soil is iron rich.

→ A greenish color indicates soil that is depleted of nutrients, such as oxidized iron, due to leaching or a lack of oxygen. Under these conditions, iron is present in its reduced form, known as ferric iron.

3 You can also study the soil texture by running your finger over the smeared sample. Can you feel crystals? This could be quartz, for example. Is there evidence of past human introduced materials such as terracotta from old drain pipes or charcoal residues from wood ash?

Determine Soil Texture

To assess a soil's texture, meaning the ratios of its mineral particles, you can use a very simple method called the sausage test or the doughnut test.

This method allows you to identify and analyze your soil texture by feel in just a few minutes. The extent to which the sample holds together will indicate its cohesion and thus its clay, silt, and sand content.

The Method

1 Take a handful of slightly moist soil as your sample.

2 Knead the sample and roll it in between your palms to form a sausage.

3 Observe your sample:

→ If you cannot roll it into a sausage because the material slips through your fingers and falls apart, the soil has a sandy texture.

Note from Jean-Martin Fortier

Ideally soil texture should be balanced: enough sand so that water can circulate, enough silt to support aggregation but not too much, and enough clay to retain nutrients. When conducting this test, roll the sample into a ball in your hands; it should hold together when you knead it. To determine if it has a loamy texture, just drop it on the ground. If the ball breaks apart on impact into coarse fragments without completely disintegrating, you have a well-balanced, loamy soil.

→ Your sample is somewhat moldable, meaning it has some plasticity, and can be rolled into a sausage. However, it cracks and tends to break when bent. This soil has a loamy texture.

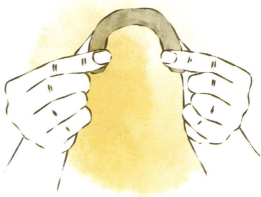

→ The sample is quite moldable, like putty, and easily forms a doughnut, for example, without any breaking or cracking. The surface may even be a little sticky. This soil has a clay texture.

Analyze Particle Size Distribution

To refine the results of the previous soil texture test and to more precisely determine the clay, silt, and sand ratios in your soil, you can perform the sedimentation test, also known as the jar or test tube test. This involves pouring soil into a glass jar so you can clearly distinguish the layers of sand, clay, and silt and then determine their percentages.

The Method

 Fill a jam jar halfway with soil scooped out of your garden.

2 Pour water into the jar, to about an inch from the top, then close it and shake vigorously for 3 to 4 minutes. Let the mixture stand for 30 minutes.

3 Shake the jar again then let it stand for 24 hours. If the mixture is still cloudy after that time, let it settle for longer and don't handle the container.

Organic matter

Clay

Silt

Sand

4 Analyze your results. Gradually the heavier sand settles to the bottom of the jar. Silt makes up the middle layer, just below the clay, which is at the top. Organic matter particles of different sizes float in the rest of the water and at its surface. This is a colloidal suspension formed by clay colloids.

Note from Jean-Martin Fortier

At the end of this test, simply measure the height of each layer and the combined layers, using a ruler (ideally in millimeters); then divide the height of each layer by the total height and multiply by 100 to get the percentage for each mineral. Next, carry these values over to a texture triangle (p. 28) to get a clearer picture of your soil's texture, still approximate but more precise than the sausage test.

Assess Soil Structure

Unlike texture, which is based on the ratio of mineral particles, soil structure refers to how they come together. Are they compacted, crumbly, or granular?

When mineral particles are well distributed, and connected through the clay-humus complex (see p. 39), with small air pockets between them, the soil is deemed to have a crumb or granular structure. This is a balanced structure. Conversely, a soil is said to have a single-grained structure when the mineral particles do not form aggregates, also called peds, due mainly to a lack or absence of binding agents (clay-humus complex).

You can run this test by digging a section of soil. After pulling the spade out of the ground, look at the amount of soil that has stuck to the steel blade. Not much soil indicates a good structure. This will tell you whether your soil is soft and loose or hard and compacted.

The Method

 1 After spading the soil, let it settle for a few days then scoop up a handful and gently press it between your hands.

2

Next, analyze its appearance.

→ The handful of soil easily forms a compact ball. When you open your hands, the soil stays in a solid clump, even when you try to pull it apart with your fingers. This soil has a compact structure. In the field, it will stick to tools you work the soil with.

→ The soil does not mold easily and has what is known as a single-grained structure, with no aggregation between the different solid particles. This soil is too permeable and does not hold water and dissolved nutrients.

→ The handful of soil sticks together but without becoming overly compacted. It holds its shape but it is still granular and easily crumbles into smaller aggregates when you open your hands. This is called a crumb or granular, and therefore balanced, soil structure, indicating satisfactory biological activity.

In the field, this soil is easy to work and will not stick to metal parts of tools, even in rainy weather.

Measuring Soil Biological Activity

Soil biological activity is proportional to the work of the pedofauna and pedoflora living within it: high levels mean healthy soil.

The organisms that work the soil are invaluable assets for improving its fertility and maintaining its health. As a result, crops have better yields and are less susceptible to disease, so it's important to regularly assess your soil's biological activity, season after season. As with texture and particle size, you can measure its biological activity with an at-home test, such as the "soil your undies" test!

Behind this surprising name lies a creative way to assess your soil's biological activity. The phrase was mostly chosen to make gardeners smile, although any fabric piece will do, provided it's cotton; being 95% cellulose, it can be broken down quickly by microorganisms. Deterioration is clearly visible, without using a microscope, so it's an impressive demonstration of what microorganisms do and why they matter.

The Method

  1 With a spade, gently create a small trench in the ground deep enough to fit the underwear, disturbing the soil structure as little as possible. Let the elastic waistband poke out or mark the location with stakes, then return the soil, without overmixing it, to cover the underwear.

2 About two months later, gently clear away the soil to expose the undies and observe their deterioration.

3 Analyze the degraded underwear:

→ Little or no sign of deterioration does not necessarily indicate that there are few or no microorganisms but may depend on factors such as low temperatures, dry weather, and lack of soil moisture. You should run the test in the spring or summer when soil biological activity is at its peak.

→ The extent to which deterioration occurs depends on your soil type and its microorganic life. After 2 to 3 months, only the elastic and stitches remain because they are thicker and take longer to break down or contain synthetic materials. However, the rest of the fabric has disappeared entirely, indicating active microorganisms and intense biological activity. Your soil is very much alive!

Test Soil pH

Hydrogen potential, or pH, is a measure of hydrogen ion activity that indicates whether a substance is acidic or basic. It can determine the acidity of a soil as well as its fertility.

Hydrogen potential (pH) is measured on a scale from 0 to 14. At the lower end, 0 indicates the highest level of acidity, while at the top of the scale, 14 is the most basic or alkaline value. In the middle, 7 represents an equal concentration of acids and bases, meaning the soil is neutral.

In home and market gardens, soils will never reach extreme pH levels; 5 is deemed highly acidic, 9 is quite basic (alkaline), and 7 is neutral.

In the soil, pH is involved in several important chemical processes and indicates the availability of essential plant nutrients. Most plants can tolerate a pH between 5.5 and 7.5, but 6.5 is ideal.

For market gardeners, getting the soil pH analyzed can achieve two purposes. First, it measures the acidity of the soil solution, which is the most important value when you are looking to determine your soil's pH. Second, buffer pH, sometimes referred to as BpH, measures the soil's ability to resist changes in pH. The lower the value, the more difficult it is to alter the pH of the soil solution.

When Soil Is Too Acidic

When soil is overly acidic (pH around 5), you can amend it with lime or any other alkaline mineral product, to obtain a soil pH, or water pH, closer to 6.5. However, if the pH buffer is quite low, liming may fail. We recommend addressing acidity in the fall after harvesting as any fluctuation in pH will stress crops. The amount of lime to add depends on its particle size and liming power. Sometimes labs will issue recommendations on this matter. Spread lime over your entire plot. If the quantity of lime needed is very high, amend the soil several times per year outside of the growing season.

When Soil Is Too Alkaline

When soils are too alkaline, a less common occurrence, there is typically no point in trying to alter the pH. The best solution is to enrich the soil with amendments to make sure plants can access all the nutrients they need. Adding organic matter and increasing soil biological activity are two other ways to decrease the soil pH.

The Method

Using a trowel, collect soil samples from about 4 inches deep (10 cm). To get the most representative samples possible, take them from about 10 randomly selected spots on the plot. Avoid aisles and areas where compost has been spread.

Mix the samples in a clean bucket until the soil is homogeneous then transfer about 9 ounces (250 g) of it into a 2-pint (1 L) jar and add 17 ounces (500 ml) of distilled water. Stir the mixture with a stick for 1 minute, then let it stand for 1 to 4 hours.

Before taking a measurement, stir the mixture again and then dip the pH meter probe into the liquid to determine the pH value, or use indicator sticks or strips sold at your local farming co-op or garden center.

Lab Tests

To corroborate the results of the field tests described above and to get more precise data, market gardeners must have soil tests done by a lab.

Whether you're starting out as a market gardener or preparing a new plot in a home vegetable garden, this type of testing helps you to understand your soil and its composition and fertility so you can prepare a soil amendment plan and develop a planting schedule for your first crops. Subsequently, regular soil analyses—possibly every year if you can afford to—will indicate any changes in the soil so you can adapt your amendment practices. After a few years of growing crops, you can test just once every five years. To find out how to get your soil analyzed by a lab,

contact your local university extension office, farming association, or an accredited lab listed online. The soil report will cover both the chemical state and physical properties of your soil: pH, mineral content, particle size, etc.

Soils samples should always be collected at the same time of year. After the last harvests in fall is best as this gives you time to prepare a soil amendment and fertilization plan. If several distinct soil types exist within your plots, it's a good idea to test them all. Soil in greenhouses should be analyzed as a separate sample.

The Method

Before collecting a sample from the area to be analyzed, it is important to learn about the protocols recommended by your chosen lab. It should provide documents to complete as well as instructions for collecting and shipping soil samples.

 Scrape a thin layer of soil off the bed surface to remove any rocks, organic waste, or residual potting mix that could affect the results. Avoid collecting soil from aisles or high-traffic zones. It's better to focus only on cultivated areas.

2 Collect the equivalent of one tablespoon of soil from about 4 to 6 inches (10–15 cm) deep, from at least 10 locations in the area you are testing.

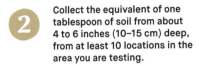

3 Combine your samples in a clean container and mix the soil thoroughly, using a tool to break up clods and obtain a uniform sample of the plot to be tested.

4 Check how much soil the lab requires then remove that amount and transfer it to the shipping bag. Mail the parcel as soon as possible.

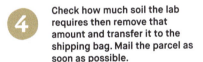

Tip from Jean-Martin Fortier

To accurately represent the area to be analyzed, you should collect soil from multiple spots, at least 10, or even more from a large plot. Sample locations should be evenly distributed throughout the area. The most effective way to achieve this without being influenced by crop conditions is to zigzag along the testing area, collecting soil samples here and there.

Identify Plant Requirements

In order to grow, develop, and bear fruit, plants need several elements: water, light, oxygen, carbon dioxide, and nutrients. It can be quite complex to manage nutrient intake as it depends on many factors, including soil type. After your soil has been tested in a lab, you can consider amending it or adding nutrients. Although plants are living beings that, like humans, breathe, eat, and reproduce, they do not express themselves, at least, not always in ways that we can perceive. Because nutritional deficiencies in plants are often difficult to identify and easy to confuse, you must exercise caution when diagnosing deficiencies and taking remedial action. However, plants may demonstrate signs of a nutrient deficiency, especially in the foliage. Certain aspects of the foliage can then serve as an alarm bell and prompt you to take action.

Diagnosing Problems

When a problem arises—perhaps a plant is looking tired—
you must first decide how serious it is by observing the
entire crop to assess its general condition.

Blossom-end rot can be the result of a calcium or magnesium deficiency.

Fungi, such as botrytis (gray mold), can also damage plants. Botrytis typically arises from too much nitrogen input during fertilization.

When assessing the state of a crop, ask yourself the following questions. Is the problem affecting multiple crops, a single crop, or one single variety? Are the symptoms affecting one isolated plant or several random plants in seemingly unrelated locations? Did the problem develop suddenly or gradually?

Identifying Possible Causes

When only one plant is affected, a nutrient deficiency may not be causing the problem. The damage could be due to a parasite or a disease with similar symptoms. To rule out this possibility, carefully inspect the foliage and, if necessary, uproot the plant to look more closely at its roots, leaves, flowers, and the inside of the stem to check for insects, larvae, and diseased plant tissue. After you can rule out damage caused by pests, fungi, viruses, and bacteria, you can consider nutrient deficiencies.

Start by analyzing the area and assessing its soil conditions and see whether events such as frost, lack of sunlight, inadequate crop maintenance, or overfertilizing could have generated these symptoms.

Nutrient Deficiencies

Deficiencies arise when a nutrient is either absent from the soil or it is present but not available to the plant.

Issues encountered in tomato crops can be attributed to various deficiencies.

Sometimes nutrients are present in the soil but are not absorbed by plants, or the plants are unable to deliver nutrients to the organs that need them. Several factors can cause this, the most common being a pH level that is not suited to the plant, inadequate irrigation, and poor drainage. At the beginning of the season, such deficiencies can also manifest if the soil is too cold. An imbalance in soil elements can also be the cause, though it is less common. Before applying any remedial treatments, you should check whether the above risk factors are causing the nutrient deficiency.

Nitrogen (N) Deficiency

This deficiency tends to occur in light, well-drained soils. To address it, simply keep the soil covered, avoid overwatering, add organic matter, let cover crops grow, and apply correct doses of fertilizers. Regular hoeing also aerates the soil and makes nitrogen available to the plants.

SYMPTOMS:
· Older leaves turn yellow.
· Growth slows down.
· Leaves are smaller.
· Older yellow leaves fall off the plant.

The first symptoms can be reversed with a nitrogen input (manure, compost, etc.). The amendment must be thoroughly incorporated into the soil and should not touch the base of each plant as it could cause fertilizer burn. When exposed to nitrogen levels that are excessive, plants grow too fast and their tissues attract aphids.

Phosphorus (P) Deficiency

With phosphorus deficiencies, the nutrient tends to be present but unavailable. This commonly occurs in the spring when the soil is still cold and saturated. A pH imbalance may also prevent plants from properly assimilating phosphorus.

SYMPTOMS:
· Leaves show a red or purple discoloration.
· Plants appear stunted, stiff.
· Few flowers and fruits.

This nutrient deficiency is uncommon when soils are amended with organic fertilizers because their phosphorus content exceeds plant requirements. It is worth noting that excess phosphorus in the soil can pollute waterways. Mycorrhizae provide invaluable support carrying phosphorus to the plants that need it.

Potassium (K) Deficiency

Insufficient potassium generates visible symptoms on the leaves and fruits, which need it to form.

SYMPTOMS:
· Leaves turn yellow, dry out, and curl along the edges.
· Brown spots may appear on the foliage.
· Yellowing between leaf veins (interveinal chlorosis).

To address this deficiency, you can add a potassium-rich organic fertilizer that contains both phosphorus and potassium. Compost is an important and affordable nutrient source to increase the potassium content of a soil. Garden waste (dead leaves or lawn clippings) as well as kitchen scraps (fruit and vegetable peels) naturally provide potassium, as do wood ash and cover crops (clover, alfalfa, vetch).

Calcium (Ca) Deficiency

This deficiency is generally caused by improper irrigation or overly rapid growth in periods of high temperatures. It is common in poorly drained or overly dry soils and soils that are acidic or potassium rich.

SYMPTOMS:
· Dying (necrosis) at the tips of leaves (especially cabbage, cauliflower, and celery).
· Necrosis at fruit ends (especially tomatoes and peppers), called blossom-end rot.

Generally there's no point in adding calcium as soils almost always hold sufficient amounts. Instead try to irrigate consistently and make sure that conditions encourage more steady plant growth. In the most serious cases, treat the leaves with calcium chloride. If needed, you can add calcium by amending the soil with lime, bonemeal, oyster shell flour, or gypsum.

Magnesium (Mg) Deficiency

This nutrient deficiency is primarily seen in greenhouse tomatoes. For prevention, we recommend adding magnesium-rich lime. To remediate this deficiency, you can apply a solution of Epsom salt, or magnesium sulfate, to young plants or the soil.

SYMPTOMS FIRST AFFECT LOWER LEAVES:

· Interveinal chlorosis, or discoloration of tissue between the leaf veins.
· Leaves dry out.

Iron (Fe) Deficiency

This nutrient deficiency is generally found in young plants grown in nurseries and particularly impacts peppers. Unlike other deficiencies, it can affect multiple plants located far apart from each other.

SYMPTOMS:

· The youngest leaves turn yellow, while veins may remain green in some crops.

It is typically caused by an overly alkaline pH and is seen in soils with low soil organic matter. Seedlings can be treated with an iron sulfate solution.

Boron (B) Deficiency

This deficiency primarily affects Brassicaceae (cabbage, cauliflower, broccoli, kohlrabi) and Chenopodiaceae (beets). Since it cannot be treated, preventive action is your best choice if the soil has a low boron content. We recommend applying the following mix once, two weeks before potting up seedlings and again a few weeks later: 20% boron solution, 1 to 2 grams per liter or quart for an area of 100 ft². Be careful, however, because excess boron can harm some crops, so we don't recommend applying full-strength undiluted boron directly to the soil.

SYMPTOMS:

· Mild deficiencies—leaves wilt.
· Severe deficiencies—stems become hollow and the plant's core may turn black.

Other Nutrient Deficiencies...

Plants also need copper (Cu), zinc (Zn), sulfur (S), manganese (Mn), and molybdenum (Mo) but in very small quantities. Deficiencies of these elements generally indicate they are unavailable for plant uptake, meaning the pH is off. Because these deficiencies have similar symptoms, they can be difficult to identify and tell apart. To remedy this, you can opt for a foliar treatment, but beware! These micronutrients can become toxic when applied in doses that are too high. The safest option is to amend the soil with compost, which largely meets all secondary nutrient needs.

The Market Gardener Method has two requirements that, at times, are in opposition and need to be managed well. On one hand, it prioritizes cultivation practices that improve soil life and encourage intense biological activity, resulting in very healthy soils. On the other hand, it requires production practices that are hyper-efficient while avoiding additional costs and wasted time. This means actively drawing upon soil nutrient reserves and soil life to optimize yields. In keeping with this philosophy, soil work—although essential—should not be overly time-consuming or require a great deal of effort. This means no plowing and no loosening the soil at depth. The keystone of the Market Gardener Method is the living soil. The more you can develop and maintain this ecosystem, the better your results. A number of strategies should therefore be implemented to conserve, nurture, and feed soil life while protecting and covering the land so that you can devote as much time and attention as possible to growing your vegetables.

Work and Improve Your Soil

Prepare Your Soil

One of the fundamental principles of the Market Gardener Method is that growers must cultivate living soil. In this context, the amendments applied to your crops matter less than the biological life of your soil. The more it thrives, the more it will have capacity to transform organic matter, thus improving soil health and fertility. In return, vegetable crops will be more robust and bountiful.

To achieve this outcome, you can implement several strategies starting in the fall. The first is to improve and maintain your soil after getting a soil analysis done. Once you identify your soil type, we recommend organizing your garden into plots, then dividing them into permanent beds to optimize the space. Over the years, your established beds will be amended and fertilized, and their soil will constantly improve, thus eliminating the need for significant tillage, preventing fertilizer and compost wastage, and increasing yields.

Tillage

Tillage is the process of turning over soil with mechanical equipment like a plow or a rotary tiller. This controversial practice should be limited to avoid disturbing soil life and harming soil organisms.

The walk-behind tractor is the only fossil fuel-powered mechanized equipment used in the Market Gardener Method.

What Is the Purpose of Tillage?

Tillage has been widely practiced because it serves several important functions during soil preparation. First, tillage loosens the soil, so that roots can grow well, then turns it over, pushing organic matter deeper into the subsoil, closer to root systems. In the process, it also breaks the surface crust created by rainfall, repeated irrigation, and trampling. This compact, firm, superficial layer prevents the exchanges of gases between the soil and the atmosphere (soil aeration) and slows down infiltration of precipitation and irrigation: it runs off rather than seeping into the ground. Lastly, tillage neutralizes weeds, which are serious adversaries for market gardeners and vegetable crops because they compete with plants, depriving them of the essential nutrients for their growth.

The Downsides of Tillage

Tillage does, however, present many drawbacks. It is particularly harmful to earthworms, especially anecic worms, which are invaluable helpers for market gardeners. The process disrupts their activity by moving them and destroying their network of burrows. Remember that these organisms are major players in soil life, keeping it vibrant and rich. Similarly, tillage degrades soil structure, leaving it more vulnerable to the effects of wind (erosion) and rain (gullying). As a result of tillage, soil will have less capacity to hold water, nutrients, and trace elements, negatively impacting vegetable growth. It also gives the false impression of improving fertility in the short term as turning over the soil provides a lot of oxygen. This speeds up decomposition in the soil organic matter, which soon becomes exhausted by this increased activity. With a lower organic matter content, the soil is then unable to generate humus naturally. As a result, soil biological activity slows down, eventually disappearing. Lastly, after repeatedly tilling, a plot begins to form a hardpan. This smooth and compact layer between the topsoil and subsoil stops air, water, and nutrients from moving freely between the layers. To break the hardpan, you need to till again, going a little deeper every time. Instead of providing a solution to the problem, tillage exacerbates it by creating a vicious cycle.

Minimum Tillage

Fortunately, it is possible to limit tillage and still support healthy soils with a high organic matter content, as well as thriving earthworms! Both professional market gardeners and home gardeners are increasingly choosing this approach. Instead of turning over the soil, they disturb it as little as possible, working only the top 8 or so inches (20 cm) to decompact and loosen garden beds. This work is done without motorized equipment, relying only on hand tools like the broadfork or the campagnole (a variation of the broadfork).[1]

The broadfork is emblematic of biointensive market gardening.

As we see it, it's not about doing away with tillage altogether, as is the case with no-till practices. The soil still needs to be loosened and decompacted but only at the surface and between crop successions to ensure decent yields. This shallow tillage also helps control weed growth, incorporate organic nutrients, and carefully prepare soil surfaces for planting or seeding.

---

1   See *Vegetable Garden Tools*, in the Grower's Guides from the Market Gardener collection.

Bio-Tillage

Soil microorganisms automatically regenerate and feed plants, creating an ecosystem that takes care of loosening the soil efficiently and naturally.

A Highly Complex Ecosystem

Organic matter is the basis of all living soils. As it is broken down by living organisms, it releases nutrients that plants can assimilate. Once this work is done, the clay-humus complex can form and actively participate in building soil structure. Air pockets then develop between soil aggregates, representing up to 60% of the soil structure. They make the ground looser and softer, which is obvious when you step on it. Microorganisms inhabit these soil pores where water and oxygen accumulate, transforming and generating nutrients.

After you spread organic matter over the soil surface with a bed preparation rake, earthworms take care of carrying it deeper into the soil.

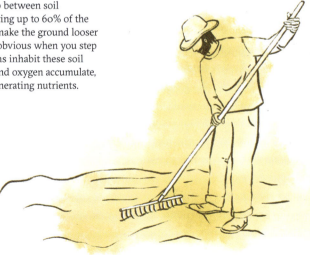

Assuming all other conditions are right, roots grow until they reach these air pockets, seeking water, oxygen, and all the nutrients that plants need to grow. To achieve this result, you don't have to incorporate organic matter deep in the soil, as growers historically have done when plowing after spreading it. Amending the surface is enough because earthworms and the soil macrofauna transport it to deeper soil layers. Over time, this can become a virtuous cycle. As the soil is more aerated and develops better structure, sometimes referred to as tilth, it supports a more diverse soil life, and vice versa!

Note from Jean-Martin Fortier

In living soils, earthworms and cover crops (see p. 89) work together. The burrows dug by earthworms and the pathways left by decomposed roots, which contribute organic matter, create a network of tunnels that loosen the soil and convey nutrients to its deeper layers. Earthworms and microorganisms also secrete a carbohydrate-based glue that helps develop the clay-humus complex.

Organize Your Garden Plots

---•---

To get the best possible yields, you must carefully consider how your plots are distributed and how beds are organized within each one.

Permanent Beds

The Market Gardener Method is based around a system that uses permanent beds, i.e., rectangular strips of soil. With 30-inch (75 cm) wide beds, you can prepare the soil, grow vegetables, care for crops, and harvest them—all without trampling the surface, thus preventing soil compaction. Many tools are manufactured for this bed size and allow growers to work particularly efficiently[1]. The beds are separated by aisles, or footpaths, at least 18 inches (45 cm) wide. To move around

and access crops, while either standing or kneeling, growers must only use the aisles. Bed length varies depending on the total available land on the farm. It tends to be 50 to 100 feet (15–30 m) long for market gardens and 5 to 30 feet (1.5–10 m) long for home gardeners. In greenhouses and high tunnels, aisles are usually narrower to maximize growing space. A width of 12 inches (30 cm) is enough to get around.

---

[1]   See *Vegetable Garden Tools*, in the Grower's Guides from the Market Gardener collection.

tunnels

tool shed

garden beds

Field blocks and garden plots, which are divided into beds, are grouped around farm buildings for market gardeners or a shed for home gardeners.

Should I Raise My Beds?

When preparing garden blocks, you could raise beds by a few inches for several reasons. Primarily, they improve drainage in the cultivated area and prevent crops from being flooded during heavy rain or spring snowmelt in northern regions. This reduces gullying and waterlogging, two processes that would, in part, undo your work to improve soil health. Soil in raised beds also warms up faster in the spring, which is a great way to start crops earlier, especially in regions with harsh winters. From a more practical viewpoint, the bed being higher than the aisle further delineates the two surfaces, making it less likely for growers to accidentally step onto a seeded bed while distracted.

Should I Seed a Cover Crop in the Aisles?

While you could seed aisles with a cover crop like fescue or crabgrass, we do not always recommend this practice for practical reasons. First, the soil from the aisles is regularly used to raise beds. A cover crop would make this operation more difficult and increases the risk of spreading the cover crop onto the bed, which is possible any time.

With aisles that are 18 inches (45 cm) wide, growers can reach the crops, right to the middle of the beds, which are raised 6 to 8 inches (15–20 cm).

Furthermore, our minimum tillage method is based on covering beds between crop successions with occultation tarps. To save time, the entire plot is covered, including the aisles, precluding any new planting or cover crop seeding. Perhaps the most interesting and least labor-intensive approach for avoiding bare soil in the aisles is to turn them into composting areas.

Simply pull crop residues and other organic matter like straw into the aisles and let them decompose. The covered aisles allow growers to keep their feet clean and limits soil compaction. Also, with the aisles lower than the beds, they are more moist than the rest of the plot and therefore more conducive to decomposing green waste and organic matter.

Tip from Jean-Martin Fortier

Raising beds with a shovel is tedious work. It's easier to use a moldboard plow in the aisle, which pushes soil out about 16 inches (40 cm) to either side of the plow toward the center of the adjacent beds. You can easily maintain a good bed height by running this tool down the aisles every 3 to 5 years, once the beds begin to settle.

Protect
Your Soil

Growers who are not regularly tilling need a simple and highly effective solution to limit the spread of weeds while improving soil biological activity. Enter mulching. Organic or synthetic, mulches create an optimal environment to initiate the humus formation process. If you recently established your farm or home garden, this technique will take some time to bear fruit, especially if your soil has a high clay content. In well-established field blocks and garden plots that have been cultivated according to the principles of biointensive market gardening, mulching the soil surface reinforces the beneficial properties of a balanced, loamy soil. It easily breaks apart when cultivated to reveal an entire network of burrows, mostly made by earthworms, which help aerate and loosen the ground. Applying a layer of leaf mulch or ramial chipped wood (RCW) in the fall is a good strategy to regenerate the soil before the next season.

Organic Mulch

―――――――――◆―――――――――

Organic mulches are made up of organic materials like straw, RCW, or compost and serve multiple purposes. They both protect the soil and improve it.

Straw

Quite popular with home and professional market gardeners, straw mulch provides many advantages. Spread over beds and between vegetable seedlings, it keeps light from reaching the soil and therefore slows weed growth, it reduces soil erosion, and it protects the surface from rainfall and irrigation that would otherwise cause compaction. Straw mulch also helps keep the soil cool and moist, and its decomposition contributes organic matter and supports microbial life. In winter, it can even be used to protect certain crops, such as root vegetables, cabbage, and leek, from frost.

The downside is that straw may contain weed seeds, pathogens, and even pests. Therefore we recommend obtaining clean straw from growers whose philosophy and farming practices you are familiar with and using it in moderation, alternating it with other mulching materials. Moreover, straw can take several months to decompose, leaving residues that may complicate some operations like seedbed prep for a crop requiring a clean surface. It's best to use straw for crops with a long growing season, like fruit-bearing vegetables.

Other Mulches

Wood chips, dead leaves, and shredded waste from hedge pruning can also be used as mulches. They are a better fit for aisles than for garden beds because of their coarse texture, making it harder to cultivate the soil, like hoeing. These mulches are also incompatible with crops that have recently been seeded or transplanted as the young seedlings require a clean, fine surface to become established. As they break down, these organic mulches stimulate fungal and microbial life in the soil while suppressing weeds. However, with coarse, woody matter, decomposition tends to be slower, and—most importantly—it has the potential

Mulching with straw has many advantages, but they do not preclude the use of other materials.

to cause a nitrogen deficiency in vegetables, sometimes called nitrogen starvation. This happens when microorganisms digesting carbon in woody mulches require nitrogen in the early stages of decomposition, depriving nearby vegetable crops of a precious resource. It's therefore best to use these organic mulches in aisles rather than on garden beds.

Ramial Chipped Wood (RCW)

RCW is a blend of shredded wood from small live, green branches, preferably from deciduous trees. First developed in Quebec in the 1970s, it made its way to Europe in the 2000s. Branches go through a wood chipper or shredder, creating a mulch to spread over bed surfaces, right up to the base of your plants. It is high in lignin, nutrients, sugars, proteins, cellulose, and tannins.

RCW is not meant to be composted. You should use it immediately, spread on the soil surface so that decomposition can continue upon contact with the soil. As it breaks down, RCW attracts and feeds fungi that in turn attract pedofauna. All these organisms develop perfect symbiotic relationships that reproduce natural life cycles occurring in forests.

There are many advantages to using RCW, including recycling plant waste after pruning hedges or trees, feeding and loosening the soil without human intervention, and maintaining soil life. However, the use of RCW can cause nitrogen starvation, which manifests as slower growth and leaf discoloration in plants. It occurs after spreading mulches such as straw, wood chips, and shredded branches (RCW, in this case), which are high in cellulose and lignin, two components of wood. The microorganisms that break down the mulch take up nitrogen, which plays a critical role in their work, thus depriving nearby plants of this nutrient source. Such mulches should therefore be used sparingly in carefully measured quantities. If you are dealing with a nitrogen deficiency, apply a fertilizer tea made from comfrey or beet molasses.

1 From November to March, collect small green twigs and young deciduous branches that are less than 3 inches (7 cm) in diameter from winter pruning of noble tree species, such as oak, chestnut, maple, beech, acacia, and hornbeam. Avoid large branches and resinous species (pine, spruce, cedar).

2 Right after shredding, apply the RCW in a 1- to 2-inch (2–5 cm) layer over your beds, then lightly rake the surface to incorporate it into the soil. In the spring, plant or seed your first crop directly into the bed after quickly leveling the surface.

Occultation

Occultation tarps (typically silage tarps) are synthetic mulches that significantly reduce the need for cultivation. They are essential on vegetable microfarms and can also be used in home gardens.

Why Cover the Soil?

Like organic mulches, occultation tarps (see additional information in *Vegetable Garden Tools* in the Grower's Guides from the Market Gardener collection) temporarily protect the soil by keeping moisture in the ground and preventing surface erosion caused by rain or waterlogging due to heavy storms or snowmelt. Under the tarp, the soil can maintain its structure and retain the nutrients that would otherwise leech into deeper layers.

If the tarp is left on for 2 to 3 weeks, weeds will grow, encouraged by the warm, moist environment at the soil surface, but eventually die due to a lack of light. This passive weeding technique is a real time-saver! These same conditions also speed up the decomposition process in residues from the previous crop.

Moreover, tarps prevent sunlight from reaching the soil, creating ideal conditions for increased soil biological activity. Earthworms, protected from their predators, can work double-time breaking down organic matter.

How to Set Up Tarps

For market gardeners, we recommend tarps that are at most 100 feet by 28 feet (30 x 8.50 m). Moving them will require several people. Home gardeners are better off with tarps that are roughly 12 to 16 feet (4–5 m) long in all directions.

1 After harvesting, if needed, use a flail mower to mulch any crop residues at the soil surface. Moisten the soil before covering it to stimulate biological activity. Spread the tarp down the length of the first bed or the edge of the field block, then unfold lengthwise, pulling on both ends to cover the plot.

2 To keep the tarp from flying away, line the edges with weighted bags, UV resistant if possible, set 5 feet (1.5 m) apart. The easiest way to fill them is to use whatever is lying around, at no extra cost: sand, rocks, or soil.

3 After a few weeks, remove the tarp in dry weather to fold and store it properly, then begin loosening the surface and get your next crop in the ground as planned.

Tip from Jean-Martin Fortier

Although tarps are not eco-friendly, they are reusable and can last up to 10 years when stored carefully. They make it possible to prep beds effectively and efficiently, and with minimal soil disturbance, which aligns with biointensive gardening principles. Tarps also keep beds relatively dry so that once uncovered, crops can be sown into them right away.

What About Woven Ground Covers?

Unlike silage tarps, woven ground covers or geotextile fabrics let rainwater through to the soil. So, before using them, you must make sure that the soil won't get waterlogged. Although woven fabric ground covers, like silage tarps, also slow weed germination and speed up decomposition in plant residues, they are a little less effective. You can use them on crops that spend a long time in the ground, like cabbage.

Solarization

While occultation tarps only warm the soil superficially,
clear plastic tarps let sun rays through and capture heat.
This is solarization.

What Is the Purpose of Solarization?

Transparent plastic amplifies solar energy and heats the soil more quickly, to a depth of 12 to 18 inches (30–45 cm). In the process, it destroys weeds, although less effectively than occultation, and kills pathogens. On the surface, temperatures can reach 140°F (60°C), so you must be careful using solarization tarps in the summer as they can harm soil microorganisms. It's best to deploy clear tarps in the spring when the sun is less intense. Instead of buying new tarps, you can salvage old plastic film from tunnels and greenhouses, which is good for your budget and the environment!

Layering Tarps

To save time in the spring, we recommend layering tarps (one clear plastic tarp and an occultation tarp), after harvesting in the fall, on the beds intended for early vegetables. Place bags filled with sand along the edges of the tarps to keep them from flying away. Once spring rolls in, simply remove the occultation tarp 1 to 2 weeks before planting, leaving the clear tarp to warm the soil while keeping it dry. The sandbags along the edges keep the tarp on the ground.

When it's time to plant your spring vegetables, you just need to pull off the solarization tarp, prepare the soil, and sow the crop. To retain the heat stored in the ground and keep the soil warmer during germination and early plant growth, you can set up a floating row cover or a low tunnel.

Tip from Jean-Martin Fortier

When rainwater pools on tarps, they become too heavy to move. You can create a slope in the tarp by pulling one corner diagonally. If needed, consider poking holes into it to allow water to drain. Rodents are another challenge with tarps, but you can limit damage by setting a few traps along the edges or—better yet—getting a cat to hunt them!

Feed and Improve Your Soil

Once you've prepared your permanent beds and the ground is loose and weed-free, it's time to amend and improve the soil. Over the years and with each harvest, soil used to grow vegetables gradually becomes depleted so you must feed it regularly. You can choose from several processes, starting with cover crops supplemented by compost. Together, they feed the soil and the organisms inhabiting it.

One final strategy can complement your strategies to enrich the soil: a crop rotation designed to avoid depleting certain soil resources and to prevent disease and pest damage. However, this practice requires meticulous planning and adhering to your schedule. Although crop rotations provide many advantages, it may be best not to implement one in your first year of operation. Waiting 2 or 3 years allows you time to get to know your soil, the farm or garden environment, and specific climate conditions before initiating a crop rotation.

Cover Crops

Cover crops, sometimes called green manures, are grown to provide soil coverage and maintain soil fertility when garden beds are not in use.

The Benefits of Cover Crops

In addition to keeping soils weed-free, growing cover crops (green manures) provides many short- and long-term benefits to the soil's health and fertility.

Protective Properties

Growing cover crops over the winter creates a lasting ground cover that protects beds from erosion caused by wind and precipitation. This practice slows down the leaching of nutrients deeper into the soil or the water table, thus maintaining soil fertility.

Soil Structure

The roots of cover crops make their way deep into the soil, loosening and opening it up like tilling does but without tools. They encourage aggregation and the development of macropores, which facilitate water and air circulation through the soil. Therefore deep-rooted crops are the best fit for green manures. They also stimulate biological activity around their roots, helping to form a clay-humus complex.

Better Biodiversity

Cover crop roots release exudates, organic fluids that nourish the soil microfauna, which in return feed the plants, creating a mutually beneficial relationship. Cover crops provide good shelter for microorganisms, protecting them from bad weather and predators. They also produce nectar and pollen that attract pollinators.

Improved Fertility

By feeding soil microorganisms, cover crops boost biological activity that makes essential nutrients available to plants. Some even pull elements from deeper layers and store these in their roots or foliage, later releasing them at the surface as they break down. This makes the nutrients more readily available for the next crop. Finally, members of the Fabaceae family (legumes) are unique cover crops. They are the only plants capable of pulling nitrogen from the atmosphere and storing it in nodules on their roots (nitrogen fixation), thus restoring soil nitrogen.

Note from Jean-Martin Fortier

The best time to mow a legume cover crop is right before it flowers. These young plants contain more nitrogen and decompose more easily, allowing you to start the next crop sooner. To fix atmospheric nitrogen, legumes require certain bacteria, rhizobia (see p. 33), that form nodules on their roots. In some soils, especially where legumes have not been grown for some time, there may be no rhizobia at all, making nitrogen fixation harder or even impossible. To determine if your garden has this problem, dig up a Fabaceae plant that is several weeks old and look at its roots. If there are no nodules, you can buy an inoculant form of rhizobia to mix into the seeds when sowing your next legumes.

Nodules are knob-like bumps that store nitrogen and form on the roots.

Choosing a Cover Crop

Many plants can be used as cover crops, including lacy phacelia, peas, white mustard, vetch, rye, red clover, buckwheat, fenugreek, and alfalfa. For spring and summer, the best strategy is to opt for a mix of legumes and grasses, such as peas and oats. Rye-vetch blends are best suited to the fall and winter as they withstand the rigors of winter and regrow well in the spring, even in northern climates. Beds where this specific winter mix has grown will not be suitable for early vegetables. Instead, plant early crops in beds that have been tarped all winter.

Broadcast seed cover crops much more densely than recommended on the packaging because the goal is to quickly get full coverage in the beds while limiting weed germination.

Cultivating Cover Crops

1 After harvesting a crop, use a flail mower, if needed, to mulch plant residues remaining on the surface. Moisten the soil before covering it to encourage biological activity. Pull the tarp down the length of the first bed or the edge of the field block, then unfold lengthwise, pulling on both ends to cover the plot.

2 After 2 to 3 weeks, uncover and broadcast seed the cover crop, then use a rake to bury them about 1 inch (2–3 cm) deep. This will keep the birds from eating all your seeds. Plan to sow right before rainy weather so the soil will stay moist, then water with a sprinkler right after seeding for better germination.

3 Once your cover crop begins to flower, about 6 to 8 weeks later, cut it back with a flail mower, then raise the beds by running a rotary plow down the aisles. Be careful—do not let cover crops go to seed in this stage as they risk turning into weeds in the soil. With legumes, mow the crop when the nodules, initially white or gray, turn pink, which means that they have fixed nitrogen.

4 Cover the crop with a silage (occultation) tarp (see p. 83) for 2 to 3 weeks, allowing microorganisms to break down the mown residues. Decomposition time depends on ambient temperature—cold weather slows down the process—and the age of the plants. Crops decay quickly when they are young and their tissue is still tender; the process is slower with older, woodier stems. Young plants contribute more nitrogen, while older crops provide carbon, however cover crops are by no means a substitute for fertilizers.

Compost

For professional vegetable growers and home gardeners, compost is an essential amendment that can serve multiple purposes. Plus, it's easy to make.

In nature, plant waste like dead leaves generates humus (organic matter), which contains nutrients that can be assimilated by roots. Market garden growers can recreate this process using crop residues and plant waste (hedge trimmings, clippings from crops grown in the aisles, raked dead leaves). This green waste can be composted in a dedicated area that is proportional to the size of the microfarm. For home gardeners, the principle is the same except that composting is limited to just a few composters or bins. Your homemade fertilizer will be 100% natural and free!

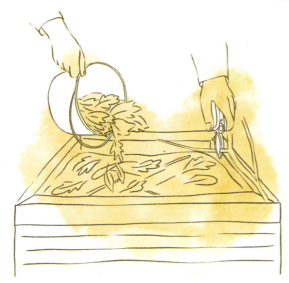

All plant waste from your garden can be composted!

Green waste undergoes several major changes to become compost, starting with mineralization (see page 36), which is a rapid transformation of soluble nutrients stored in plant residues. The humus formation process follows, initiated when nitrogen (green waste), water, and air are all present. To finish, a multitude of organisms, such as earthworms and bacteria, take over and convert the plant residues into organic matter. This process can be done without covering the pile (cold composting), producing usable compost after 6 to 8 months. However, when covered with a tarp (hot composting), decomposition takes only 4 to 6 months.

Collecting Green Waste

The materials you select must be healthy and biodegradable, and be free of pests, disease, and pesticides. Therefore it's essential to sort your waste into categories.

Organic garden waste includes dead leaves, lawn clippings, and wilted flowers but not perennial and annual weeds, like bindweed and dandelion, which are likely to propagate through seeds or root fragments. Do not add leaves from fruit-bearing trees and roses as they are potential carriers of diseases and overwintering forms of pests. Twigs, branches, and thick leaves (plane trees, laurels, and conifers) should be shredded or cut up with pruning shears to accelerate the decomposition process. Household organic waste includes food scraps (fruit peels and vegetable tops, coffee grounds, tea, egg shells, stale bread) as well as cardboard, wood ash, and pet hair. These break down quickly and provide a lot of nutrients. However, vegetable tops and peels must be combined with sawdust, ash, or wood chips to absorb the large amounts of water they contain.

Vegetable peels and tops are composted with fibrous materials (wood chips or sawdust) that can absorb excess water.

Lastly, some materials cannot be composted because they do not break down easily. This is the case for meat waste and vegetable oils, as well as waste from varnished, treated, or glued wood and newspapers and magazines that feature unspecified types of coatings and inks. Cat litter and certain fruits covered in preservatives, such as citrus, are not suitable for composting.

The Result

At the end of the decomposition process, you will have two types of organic matter. The first is mulch, a mature but still coarse compost. It can be used to cover the soil or plant new bushes—it is then considered a fertilizer—and it can simply be incorporated into garden beds in the fall along with manure. The second is sifted compost, a rich and balanced growing medium (like a potting mix) used for seeding into trays and potting up seedlings.

After sifting your compost, you will have a rich substrate to incorporate into your garden beds.

The Benefits of Compost

NUTRIENT-RICH SOIL
Rich in nutrients such as nitrogen, phosphorus, and potassium, compost supports plant growth and feeds the soil.

IMPROVED SOIL STRUCTURE
By adding organic matter to the soil, compost improves its structure and ability to hold water and nutrients while helping plant roots grow.

DISEASE PREVENTION
Compost supports soil microorganisms that prevent diseases affecting plants and keep the soil healthy.

WASTE REDUCTION
Making and using compost is also an ecological and socially responsible choice.

How to Use Compost

IN SOIL PREPARATION
Before sowing or planting a crop, amend the soil with mature compost (even if it's slightly coarse, containing pieces of branches or twigs). This improves the soil structure and enhances its nutrient content.

AT THE SOIL SURFACE
Throughout the growing season, amend beds to improve soil fertility and to promote plant growth. However, be careful not to overfertilize crops, which may cause root burn. The best way to make sure you apply the right amount of compost is to monitor soil pH by testing it regularly (see p. 54).

AS A MULCH
Coarse compost, not necessarily fully mature, can be spread as a mulch at the base of plants to help the soil retain water and nutrients. The compost will continue to decompose once it comes in contact with the soil.

IN COMPOST TEA
Compost tea, a nutrient-rich liquid fertilizer, can be made by filling a canvas bag with compost and soaking it in water for several hours. Use the liquid to water crops and apply the maceration residue to the soil as it provides nutrients that can quickly be absorbed by the roots when dissolved in water.

How to Make Your Own Compost

The process is identical for professional growers and home gardeners, the only difference being the volume of compost to be stirred, the required surface area, and the time to turn over the compost.

1 Choose a flat surface that is in contact with the soil and pile your materials in successive 8-inch (20 cm) layers. Alternate dry, or brown, materials (shredded branches, dead leaves, straw, etc.) and wet green waste (grass clippings, foliage). Depending on the amount to be composted, you may require multiple bins or piles. Water the layers evenly to keep them moist and sustain the organisms that break down plant residues.

2 Stir the compost regularly, turning it over at least once a month to thoroughly mix the layers. You may notice an increase in temperature: this is useful for speeding up the process. You can cover the compost pile with a tarp but always make sure to allow some airflow.

3 The resulting coarse compost will still contain some twigs and pieces of wood. You can use it as an amendment on your beds or sieve it to get compost soil. This fine, rich substrate can be used like potting mix for seeding into trays and plug flats or potting up seedlings.

Fresh nettles in the form of a nettle maceration residue (compost tea), are an excellent compost activator.

Tip from Jean-Martin Fortier

Avoid adding weeds and perennial plants that continue to grow even after being composted. To activate aerobic bacteria, the compost must be watered and stirred regularly. Other strategies can speed up the decomposition process, especially using a compost activator:
· Apply ammonium sulfate (nitrogen), sold commercially as a compost activator.
· Apply nettle, comfrey, horsetail leaves, or nitrogen-rich algae, used fresh or as an extract.
· Apply urine!

Crop Rotation

―――――――○―――――――

Crop rotation is the practice of planning seeding and planting from one year to the next so that the same crop species is never grown in the same bed (or in the same place).

Why Establish a Crop Rotation?

In organic farming, this ancestral practice is essential to maintain soil health and naturally prevent damage by pests and diseases. Crop rotation—changing the types of crops grown in a bed from year to year—should not be confused with crop succession, which involves planting the same crops one after the other in the same bed throughout the year. Implementing a crop rotation is quite complex, requiring you to carefully plan your market garden or home garden.

Preventing Pest and Disease Damage

By varying crops grown in a given bed, you prevent pests like nematodes and diseases like clubroot from getting established. If you avoid repeatedly growing the same type of crop in the same location, you will remove host plants from one year to the next and considerably reduce the spread of disease and pest infestations.

Improving Soil Health

The roots of all vegetables develop and grow in different ways, reaching distinct soil depths carrying various nutrients with them. Each species lives symbiotically with an array of soil microorganisms. The combined effect of this crop diversity in a single bed enhances soil microbial life and develops a rich, resilient ecosystem that more effectively resists the proliferation of pathogens.

Optimizing Amendments

Not all nutrients provided by organic soil amendments are available in the first year after their application. You can rely on this reality to improve your use of amendments and gradually distribute nutrients over several years by simply planning your crop rotation according to plant requirements. For example, after feeding your soil with manure and planting tomatoes in the first year, they will benefit from a large amount of nutrients that they need to grow (tomatoes are heavy feeders). Then in the second year, without further amending the bed, you can opt for a less demanding crop, like lettuce, or a root vegetable (carrot) whose deep roots seek out nutrients that leached into lower soil layers. Thus, with crop rotations, you avoid amending the soil every year, only adding organic inputs every 2 or 3 years. It's a great way to save money!

Instructions for Market Gardeners

Plan your crop rotation before your growing season begins. To start, draw out a preliminary plan by hand or on your computer. Professional market gardeners, and anyone managing large areas and quantities of vegetables, may choose to use specialized software and crop management spreadsheets.

Grouping Vegetables

Classic crop rotations categorize crops according to vegetable type. These include fruiting vegetables (tomato, zucchini, eggplant, pumpkin, cucumber, etc.), legumes (beans, peas, etc.), cover crops, root vegetables (carrot, beet, turnip, radish, potato), and leafy vegetables (cabbage, lettuce, spinach, etc.). New ways of classifying crops have been gradually introduced, like grouping vegetables by their botanical family, growing needs, or nutritional requirements.

Here is one way to organize crops:

1. Undemanding crops: leafy vegetables (salads) or legumes (Fabaceae). They take up garden space for up to 60 days.
2. Moderately demanding vegetables: root vegetables and alliums (onions, etc.). They take up garden space for 60 to 70 days.
3. Demanding vegetables: fruiting vegetables (Solanaceae, cucurbits, etc.). They take up garden space for more than 70 days.

4. Cover crops.
Next, assign a color to each crop category.

Assign Crop Categories to Garden Plots

After grouping all your permanent beds into several plots, your next step is planning the order of crop categories to be planted from one year to the next within each block. When doing this, make sure to follow these rules: 1) an undemanding crop always follows a demanding crop; 2) a demanding or moderately demanding crop always follows a cover crop; 3) all crops should meet your vegetable needs (for example, plan more undemanding crops if you intend to grow mostly lettuces).

Make a Detailed Crop Plan

Now that you've figured out the layout of your different plots, assign each one a crop color for the current year. Next, fine-tune the garden by choosing which vegetables to grow, the number of beds they will occupy, and the start and end dates. You can plan crop successions in these beds provided the chosen vegetables match the category you've assigned to each space.

On a microfarm, each plot should be assigned a specific crop category to make sure the same type of vegetable is never sown or planted in any given space twice in a row.

Home gardens can be split into four to apply crop rotation principles.

Instructions for Home Gardeners

In home gardens, the crop rotation concept should be organized around a classic vegetable rotation based on categories. Remember, the categories are fruiting vegetables (tomato, zucchini, eggplant, pumpkin, cucumber), legumes (beans, peas), root vegetables (carrot, beet, turnip, radish, potato), and leafy vegetables (cabbage, salads, spinach). Whatever size your garden space is, divide it into four equal plots. Then plant these categories of vegetables in each plot, rotating them clockwise, for example, from one year to the next.

Plot 1: The first year, after amending soil with manure (only a quarter of the vegetable garden will be amended with organic matter in any given year), plant fruit-bearing vegetables because they require lots of organic matter. Plot 2: Plant legumes and a cover crop that will fill the bed while continuing to feed and improve the soil. Plot 3: Plant leafy vegetables, which require little organic matter. Plot 4: Plant root vegetables, which can grow deep roots to find nutrients.

In the second year, amend plot 2 with the organic matter (manure) and plant fruit-bearing vegetables. Root vegetables will grow in plot 1, with legumes and the cover crop in plot 3. Plot 4 will grow leafy vegetables. And so on through a 4-year cycle.

Acknowledgments from Jean-Martin Fortier

I wish to thank the entire team at the Market Gardener Institute for encouraging me to pursue my mission, every day. A big thank-you also goes out to the Growers & Co. team, who pushes me to come up with new types of equipment! I especially want to acknowledge my partner, Maude-Hélène Desroches, who is an exceptional market gardener and a dear friend!

Acknowledgments from New Society Publishers

We extend a great thanks to Delachaux et Niestlé, the French publisher, for working with us to publish this English edition. Further thanks to the New Society Publishers team for producing the book and especially to Laurie Bennett for her meticulous attention to technical details and high-quality translation into English.

Acknowledgments from Delachaux et Niestlé

A big thank-you to Jean-Martin Fortier and his team at the Market Gardener Institute for this wonderful collaboration.

Our heartfelt thanks go out to Pierre Nessmann for putting us in touch with Jean-Martin, for thoroughly editing this collection, and for being so generous with his time. For this book, we owe him so much. We also wish to thank Flore Avram, whose illustrations give this collection a beautiful, simple character; to Grégory Bricout for graphic design that cleverly reflects Jean-Martin Fortier's spirit; and to Sandrine Harbonnier and Sabine Kuentz for their work on the text.

Reference Books

The Market Gardener Masterclass,
www.themarketgardener.com
The Market Gardener: A Successful Grower's Handbook for Small-Scale Organic Farming, New Society Publishers, 2014.
Winter Market Gardening: A Successful Grower's Handbook for Year-Round Harvests, New Society Publishers, 2023.
Microfarms: Organic Market Gardening on a Human Scale, New Society Publishers, 2024.

Grower's Guides from the Market Gardener

Tomatoes: A Grower's Guide
Vegetable Garden Tools: A Grower's Guide
Root Vegetables: A Grower's Guide
Living Soil: A Grower's Guide

Coming Soon

Fruiting Vegetables: A Grower's Guide
The Well-Planned Vegetable Garden
Fall and Winter Vegetables
Produce Your Own Plants
Salads and Leafy Greens
Herbs
Perennial Vegetables

Translator: **Laurie Bennett**

About New Society Publishers

New Society Publishers is an activist, solutions-oriented publisher focused on publishing books to build a more just and sustainable future. Our books offer tips, tools, and insights from leading experts in a wide range of areas.

We're proud to hold to the highest environmental and social standards of any publisher in North America. When you buy New Society books, you are part of the solution!

At New Society Publishers, we care deeply about *what* we publish — but also about *how* we do business.

- This book is printed on **100% post-consumer recycled paper**, processed chlorine-free, with low-VOC vegetable-based inks (since 2002)
- Our corporate structure is an innovative employee shareholder agreement, so we're one-third employee-owned (since 2015)
- We've created a Statement of Ethics (2021). The intent of this Statement is to act as a framework to guide our actions and facilitate feedback for continuous improvement of our work
- We're carbon-neutral (since 2006)
- We're certified as a B Corporation (since 2016)
- We're Signatories to the UN's Sustainable Development Goals (SDG) Publishers Compact (2020–2030, the Decade of Action)

To download our full catalog, sign up for our quarterly newsletter, and to learn more about New Society Publishers, please visit newsociety.com.

ENVIRONMENTAL BENEFITS STATEMENT

New Society Publishers saved the following resources by printing the pages of this book on chlorine free paper made with 100% post-consumer waste.

TREES	WATER	ENERGY	SOLID WASTE	GREENHOUSE GASES
13	1,000	5	43	5,470
FULLY GROWN	GALLONS	MILLION BTUs	POUNDS	POUNDS

Environmental impact estimates were made using the Environmental Paper Network Paper Calculator 4.0. For more information visit www.papercalculator.org

MIX
Paper | Supporting responsible forestry
FSC® C016245

new society
PUBLISHERS
www.newsociety.com